Knowledge Coordination

Knowledge Coordination

Flávio Soares Corrêa da Silva

Jaume Agustí-Cullell

WILEY

Copyright © 2003 John Wiley & Sons Ltd, The Atrium, Southern Gate, Chichester,
West Sussex PO19 8SQ, England

Telephone (+44) 1243 779777

Email (for orders and customer service enquiries): cs-books@wiley.co.uk
Visit our Home Page on www.wileyeurope.com or www.wiley.com

All Rights Reserved. No part of this publication may be reproduced, stored in a retrieval system or transmitted in any form or by any means, electronic, mechanical, photocopying, recording, scanning or otherwise, except under the terms of the Copyright, Designs and Patents Act 1988 or under the terms of a licence issued by the Copyright Licensing Agency Ltd, 90 Tottenham Court Road, London W1T 4LP, UK, without the permission in writing of the Publisher. Requests to the Publisher should be addressed to the Permissions Department, John Wiley & Sons Ltd, The Atrium, Southern Gate, Chichester, West Sussex PO19 8SQ England, or emailed to permreq@wiley.co.uk, or faxed to (+44) 1243 770620.

This publication is designed to provide accurate and authoritative information in regard to the subject matter covered. It is sold on the understanding that the Publisher is not engaged in rendering professional services. If professional advice or other expert assistance is required, the services of a competent professional should be sought.

Other Wiley Editorial Offices

John Wiley & Sons Inc., 111 River Street, Hoboken, NJ 07030, USA

Jossey-Bass, 989 Market Street, San Francisco, CA 94103-1741, USA

Wiley-VCH Verlag GmbH, Boschstr. 12, D-69469 Weinheim, Germany

John Wiley & Sons Australia Ltd, 33 Park Road, Milton, Queensland 4064, Australia

John Wiley & Sons (Asia) Pte Ltd, 2 Clementi Loop #02-01, Jin Xing Distripark, Singapore 129809

John Wiley & Sons Canada Ltd, 22 Worcester Road, Etobicoke, Ontario, Canada M9W 1L1

Wiley also publishes its books in a variety of electronic formats. Some content that appears in print may not be available in electronic books.

British Library Cataloguing in Publication Data

A catalogue record for this book is available from the British Library

ISBN 0-470-85832-X

Project management by Originator, Gt Yarmouth, Norfolk (typeset in 10/12pt Palatino)
Printed and bound in Great Britain by Antony Rowe, Chippenham, Wiltshire
This book is printed on acid-free paper responsibly manufactured from sustainable forestry in which at least two trees are planted for each one used for paper production.

Contents

Foreword ix

Acknowledgments xi

List of Figures xiii

List of Tables xv

1 Introduction 1
   1.1 Management, Engineering and Society 2
   1.2 Goals and Motivations of this Book 10
   1.3 Intended Audience 13
   1.4 Overview of Chapters 13
       References 14

2 Knowledge 17
   2.1 Knowledge and Management 20
       2.1.1 Knowledge and the classical school of management 20
           2.1.1.1 Frederick W. Taylor and Henry Ford 21

		2.1.1.2 Henri Fayol and Max Weber	24
		2.1.1.3 Chester Irving Barnard	28
	2.1.2	Knowledge and the socio-technical school of management	34
2.2	So What Is Knowledge and Where Can We Find It?		39
	2.2.1	Knowledge as special information	44
	2.2.2	Knowledge as justified true belief	50
	2.2.3	Knowledge as status of an agency	52
	2.2.4	Knowledge as the skill to provide meaning to data	57
	2.2.5	Knowledge as the capability to change the world	58
	2.2.6	Knowledge and agencies	60
	References		64

3 Agents — 67

3.1	Agents for Knowledge Modelling	68
3.2	Agents for Organizational Modelling and Design	89
	3.2.1 Agencies and knowledge in the different schools of management	90
	References	94

4 Ontologies — 97

4.1	Ontologies – Natural and Artificial	99
4.2	Implementing and Using Artificial Ontologies	101
4.3	Illustrative Example I: The Resources–Events–Agents Enterprise Ontology	106
4.4	Illustrative Example II: The National Academic CVs Database in Brazil – Lattes	120
4.5	Natural Ontologies and Knowledge Coordination	129
	References	131

5	**Capabilities**	135
	5.1 Managing Capabilities	138
	5.2 Structures of Capability Providers	144
	5.3 Examples	146
	5.3.1 Mobile robots	146
	5.3.2 Conference speakers	153
	5.3.3 Other examples	154
	5.4 Assessing Knowledge Coordination	154
	5.4.1 Minimize $\frac{\text{(delegations)}}{\text{(task)}}$	155
	5.4.2 Minimize $\frac{\text{(agents)}}{\text{(task)}}$	155
	5.4.3 Maximize probability of cross-delegation of tasks	155
	References	156
6	**Conclusion**	157
	Bibliography	163
	Index	171

Foreword

Knowledge management became a marketing term only a few years ago – the most obvious symptom being a rush of books bearing this slogan on the bookstalls at airports. The excitement at the time came from the "soft" side of systems engineering and management analysis. This emphasized the "management" in knowledge management, and it is true that managing of knowledge as an asset (corporate, individual, social) is a major issue for everyone, now that we have access to so much of it in so many electronic forms. The management of such knowledge, however, is neither a subdiscipline of management science nor of computer science. Its true foundations are in the science and theory of knowledge representation and reasoning. For those working on those foundations, the coordination of knowledge is not a management fad, but an aspiration of those who rely on acquiring, adapting and decommissioning knowledge that they have felt the need to express in a formal way. Hence, I believe, the title of this book.

It is hard to write a well-balanced book on knowledge coordination. On the one hand, one must rely on mathematical principles that are timeless because they do not rely on the details of any particular application. On the other hand, management of knowledge relies for its effectiveness on the pragmatic methods developed specifically for applications. The book one wants must combine the mathematical and human elements of this equation. The authors of this book have attempted to achieve this balance. If you, the reader, get your monthly paycheck from coordinating knowledge, then I imagine you will meet many concepts in this book that aren't just unfamiliar, but are from a more abstract realm than the one you normally inhabit. Enjoy that experience and, if the maths feels a bit dry sometimes, then remember that it is this sort of maths that makes formal knowledge coordination possible. If, on the other hand, you are a mathematician, then you may be surprised that so much of the human side of organizational management found its way in among the equations. Bear in mind that this is the way of the world. All readers of this book will, I expect, be left with the impression that much remains unsaid. Part of this is authorial restraint, but also much of the terrain remains to be travelled. This book should help you in your journey.

David Stuart Robinson
Director
Centre for Intelligent Systems and Their Applications
School of Informatics
University of Edinburgh

Acknowledgments

The authors benefited from the following grants, which partially helped to fund the development of this work:

- the international project DECAFF-KB (Distributed Environments for Cooperation Among Formalisms For Knowledge Based Systems), funded by CAPES[1]/Brazil and the British Council between 1998 and 2000;

- the international project RC-CAI (Redes de Capacidades para Cooperação entre Agentes Inteligentes), funded by CAPES/Brazil and the MECD[2]/Spain since 2003;

- grant ACI 99-26, provided by the Generalitat de Catalunya.

Flávio thanks the University of São Paulo, especially the Department of Computer Science, for the institutional support to develop this work. He also warmly thanks his wife Renata and his daughter Maria Clara for giving him the incentive to write this book. He also thanks Iracema (hi, mum!), Yára, Paulo, André

---

[1] CAPES = Coordenação do Aperfeiçoamento de Pessoal do Ensino Superior – Federal Council for the Preparation of Higher Education Personnel.
[2] MECD = Ministerio de Educación, Cultura y Deporte – Spanish Ministry of Education, Culture and Sports.

and Evelin for their love and support. Finally, he wants to thank his father (Big Flávio) for getting him started on the road that has taken him up to here so far. He hopes he can make his dad proud of him "from up there".

Jaume thanks the CSIC[3] for general provision of the infrastructure that allowed the development of this work. He also thanks his colleagues at the IIIA[4] for such a warm and pleasant working environment. Finally, he thanks his family, who were his unending source of inspiration for carrying out work that, after all, is done on their behalf.

---

[3] CSIC = Consejo Superior de Investigaciones Científicas – Spanish Higher Council for Scientific Research.
[4] IIIA = Institut d'Investigació en Intel.ligència Artificial – Artificial Intelligence Research Institute.

Figures

2.1	A simple hierarchy – organogram.	30
2.2	Polymorphic hierarchies: (a) polymorphic hierarchy; (b) matrix structure representing a polymorphic hierarchy.	31
2.3	Network structure.	32
2.4	Total versus partial interpretations.	46
2.5	Complete versus incomplete interpretations.	47
2.6	Mappings: bijective; redundant and consistent; inconsistent and non-redundant; and inconsistent and redundant.	49
2.7	Belief of an agent.	54
2.8	Knowledge of an agent.	55
3.1	Unification algorithm.	78
3.2	Answering a query using program clauses.	79
3.3	Resolution.	80
4.1	The initial screen of Protégé-2000.	111
4.2	The classes in the REA artificial Ontology.	112
4.3	A car rental instantiation of the REA artificial ontology.	115
4.4	The REA artificial ontology – a simplified view.	118
4.5	The REA artificial ontology – excerpts from the car rental instantiation.	119
4.6	A class hierarchy for a CV.	123

4.7	A class hierarchy for a CV in Protégé-2000.	124
4.8	Slots for CVs in Protégé-2000.	125
4.9	Creating an instance in Protégé-2000.	126
4.10	HTML presentation of a class generated by Protégé-2000.	127
4.11	HTML presentation of the instance of a class generated by Protégé-2000.	128
5.1	Sequential decomposition.	139
5.2	Parallel decomposition.	140
5.3	Conditional decomposition.	140
5.4	Iteration.	141
5.5	Procedure to select a capability provider.	143
5.6	Mobile robots.	148
5.7	Partial order of robots by speed (estimated quality of service).	150

Tables

5.1	Values of attributes of robots.	147
5.2	Relevant agencies.	151
5.3	Values of attributes of conference speakers.	153

Tables

1

Introduction

Escribí este libro, sobre todo, para educarme. Intenté publicarlo porque pienso que hay muchos (...) que se encuentran en el mismo estado de ignorancia en el que yo mismo me hallaba al empezar a escribir. – D. Harvey

Providing a clear and precise definition of "knowledge" is not an easy task. It has kept Western philosophers busy throughout history – from pre-socratic philosophers to contemporary ones. We do find many definitions of knowledge, proposed by different scholars at different moments in history and based on different philosophical traditions, but these definitions are far from converging to a single, encompassing understanding of the meaning of the word *knowledge*.

Not having a single coherent definition of knowledge should not be regarded as a failure of our philosophical tradition. Indeed, the pursuit of this definition has resulted in a better understanding of humanity, society and nature, and how these three relate to each other.

Curiously, however, *not* having a single coherent definition of knowledge espoused by philosophers has not restrained businessmen from performing *knowledge management*, nor computer

---

Knowledge Coordination F. S. Corrêa da Silva and J. Agustí-Cullell
© 2003 John Wiley & Sons, Ltd ISBN: 0-470-85832-X

scientists from practising *knowledge engineering*, nor sociologists, management gurus, economists and grand organizations such as UNESCO and the EU from stating that we now live in a *knowledge society*. Presumably, *some* clear understanding of what is meant by knowledge is assumed in each of these cases (or, most likely, in each instance of each of these cases), but we feel it safe to also assume that there is not a single and coherent understanding of the meaning of knowledge for *all* these cases.

In the present book we do not address the problem of defining knowledge as philosophers. We simply analyse the underlying understandings of this word in the contexts of knowledge management, knowledge engineering and the knowledge society. By making the differences and commonalities among these understandings clear, we hope to contribute to the prevention of misunderstandings and to make room for a better flow of information and technologies among these areas.

A common central problem in these areas is *knowledge coordination*. We devote this introductory chapter to explaining why we think this way, thus justifying our choice for the name of this book. We hold back our discussion about the meaning of knowledge until Chapter 2, and for the moment we rely on the reader's own intuitions about what is meant by *knowledge* here. We have organized the book this way precisely to provoke the reader into formulating a personal definition (or some personal definitions) of knowledge, in the specific contexts mentioned above, which can be compared with our analyses and viewpoints further on.

1.1 MANAGEMENT, ENGINEERING AND SOCIETY

There are a number of definitions for knowledge management. In Tiwana (1999), for example, we find it "defined"[1] as:

---
[1] Strictly speaking, this method should never be accepted as a definition, since it uses the terms ("management" and "knowledge") to be defined as part of the definition.

1.1 MANAGEMENT, ENGINEERING AND SOCIETY

management of organisational knowledge for creating business value and generating a competitive advantage.

A clear understanding of this statement requires the concept of *organizations*, so that we know what is meant by *organizational knowledge*.

In Marshall (1999) we find some definitions of "organization":

An organisation is a consciously coordinated social entity, with a relatively identifiable boundary, that functions on a relatively continuous basis to achieve a common goal or set of goals. (Robbins, 1990)

An organisation structure is a defined set of role relationships which, implicitly or explicitly, set limits of behaviour and action and, hence, imply freedom of behaviour within those limits. Remaining within those limits ensures tranquil role interactions; conflict comes from pushing beyond the limits. (Howard, 1996)

An organisation is a network of interacting agents that create, maintain and terminate commitments. (Verharen, 1997)

We also find a definition of "organization" in Fox et al. (1998):

We consider an organisation to be a set of constraints on the activities performed by agents.

Finally, we wish to cite Barnard (1938), where we find the definition of a (formal) "organization" as:

... a system of consciously coordinated activities or forces of two or more persons.

These definitions originate from distinct viewpoints, ranging from systems engineering to sociology. They complement each other by giving emphasis to different facets of organizations.

In all definitions, we have an organization as a single entity, composed of other entities (sometimes called *agents*). Organizations can, however, be composed of other organizations. We propose *agency* as an encompassing term to describe an organization. An agency is a single entity, recursively composed of other agencies. An atomic agency is given the special name of *agent*. The distinctive property of an agent is that of being an agency that has no other agencies as components.

The components of an agency interact with each other. This interaction is regulated by explicit rules of behaviour and action, as well as by mutual commitments that are set up and terminated dynamically. An agency has a goal or set of goals, and the behaviour and actions of its components are coordinated in order to achieve this goal or set of goals.

Hence, *coordination* is at the heart of the concept of an organization, together with the concepts of agent and agency (we actually take the concept of agency as interchangeable with the concept of organization).

In the *Merriam-Webster Collegiate English Dictionary* we find the following for "management":

1. *the conducting or supervising of something (as a business)*;

2. *judicious use of means to accomplish an end.*

Organizational knowledge must be, therefore, knowledge ascribed to an agency, which in turn must be somehow associated to its components. Management of organizational knowledge is the judicious use of that knowledge to achieve some goal or set of goals. There cannot be knowledge management without coordination of the knowledge of the components of an agency.

Peter Drucker (1988) proposed that (as highlighted in Scholtz, 2002) the agencies of the future will be *coordinated* rather than managed, with knowledge disseminated across the whole agency, and the traditional command-and-control agencies replaced by information-based agencies (employing the terminology found in Drucker, 1988), based on coordination of the activities of highly autonomous entities.

In Drucker (1988) we find some interesting, proposed prototypes for the agencies of the future: companies will resemble universities, hospitals and symphony orchestras. In these three prototypical agencies, knowledge and responsibility are disseminated across the whole agency. Coordination can be centralized (as occurs, for example, in a symphony orchestra, where we have the figure of the conductor as a general coordinator), but control is distributed and we find high levels of autonomy at all levels of the agency.

The components of these agencies are self-motivated and resourceful. The success of such agencies depends on the alignment of the goals of the agency and its components. Motivation of the components of an agency therefore becomes a central concern for the success of the agency. To better understand the meaning of this, consider the motivations, autonomy and behaviour of academic researchers, medical doctors and orchestral musicians – prototypical agents of an information-based agency – and compare them with those of, for example, bank clerks and production line workers in manufacturing industries – who are in turn the prototypical agents of traditional command-and-control agencies. Self-motivation and autonomy are highly desirable or even necessary for the former, while negligible or even undesirable for the latter.

Resorting again to the *Merriam-Webster Collegiate English Dictionary*, we find for "coordination":

the harmonious functioning of parts for effective results.

Coordination of organizational knowledge is, therefore, the harmonious "functioning" of the knowledge of the components of an agency to effectively accomplish its ends.

According to Drucker (1988) the "typical manufacturing company" is the one we find "circa 1950" in the United States, and information-based agencies are starting up today. Agencies, in the sense of Drucker (1988) (and of Binney, 2002), have humans as their foundational agents. It does not make sense to talk about the engineering of the knowledge of agents, if these agents are human.

Software-based agencies are agencies in which the foundational agents are not human agents, but software agents. Software-based agencies were first inspired by human agencies. As we find in Watt (1990), the first important, high-level, *imperative* programming languages – Fortran and Cobol – also appeared circa 1950. Imperative programming mimics classical management agencies. Purely imperative programs are based on a command-and-control organization strategy: computer commands are organized as procedures that change the values of public variables held in storage, and these procedures are structured as a simple hierarchy, in which we can always identify a main procedure as the single entry point to activate the program. The main program works as a software-based "top manager" that delegates tasks to low-level procedures following a rigid hierarchy.

Some 20 years after the inception of imperative programming, object technologies appeared, and since then they have gradually replaced the command-and-control organization of programs. Objects are autonomous entities with internal and private data representations. Objects communicate with each other to delegate subtasks to each other. Hence, there is no rigid and fixed hierarchy controlling the collective behaviour of objects. A network of objects entails opportunistic, goal-driven cooperation to solve specific tasks. Object-based computer systems are the software-based counterpart to the agencies of the future proposed by Drucker.

It should become clear from this evolution of agencies – either human or software-based – that there is a shift of interest from processing to interaction and communication. As suggested in Winograd (1997), the effectiveness of modern agencies results from how processing units coordinate their work with each other, more than how they perform their work internally. By specifically considering software and computer-based agencies, in Winograd (1997) we find that:

> ... there will always be a need for machinery and a need for software that runs the machinery, but as the industry matures, these dimensions will take on the character of commodities, while the industry-creating innovations will be in what the hardware and software allow us to communicate.

Replace computer-based agencies by human-based ones, and you will find the tenets and propositions of socio-technical organizational design and semi-autonomous groups, which – as discussed in Chapter 2 – embody many of Drucker's propositions.

It does make sense to talk about the engineering of the knowledge of a software agent. Hence, if all or some of the foundational agents of an agency are software agents, we have room for *knowledge engineering*.

In Sowa (2000) we find that engineering is characterized by the use of science and mathematics for the purpose of solving practical problems. Knowledge engineering, in that book, is defined as the branch of engineering that analyses knowledge about some subject and transforms it to a computable form for some purpose.

The purpose of embodying knowledge in software agencies is to prepare them to become components of other agencies. Hence, the end result of knowledge engineering – namely, software agencies endowed with knowledge – are software agencies whose capabilities can be coordinated with those of other agencies.

In the *Merriam-Webster Collegiate English Dictionary* we find for "society":

> *an enduring and cooperating group whose members have developed organized patterns of relationships through interaction with one another.*

The knowledge society should therefore be a group whose members interact based on their knowledge.

In this book we deal with knowledge *coordination*. We take agencies comprised of human agents as well as of software-based agents into account. As thoroughly analysed and discussed throughout the text, software agents are designed agents, whereas human agents are evolving and self-designed. Theories of software agents are normative theories, whereas theories of human agents are essentially descriptive. During design, we can determine the knowledge, beliefs, goals, motivations and capabilities of software agents, but we cannot do the same for human agents. Rather, we must infer these features of human agents from their observable behaviour, and we can always be certain to have only a partial picture of them for human agents. Management and coordination theories for software agents are also designed and predetermined, and therefore can be predictable. As for human agents, they must be based on motivation theories and can at their best increase the likelihood of desired behaviours.

All sorts of knowledge, which we are going to discuss in Chapter 2, can be classified in two types:

1. knowledge that can be extracted from agencies and encoded using some sort of language, thus becoming available for further utilization independently of their originating agencies; and

2. knowledge that defies extraction from the originating

agencies, thus becoming available only through its "carriers".

The first type of knowledge is encyclopaedic knowledge. It can be exported from an agency to some neutral repository (a "knowledge base"[2]), to be later imported into another agency without the direct intervention of the first agency. Its main characteristic is the fact that it can be written down using some sort of language.

The second type of knowledge is actionable knowledge. Most typically, it is the capability of an agency to do something that can be relied upon, but whose detailed internal procedures cannot be fully explained.

In Hansen et al. (1999) it is observed that large corporations usually rely almost entirely on one of these two types of knowledge to build their knowledge management infrastructure. In Daft (2001) these infrastructures are respectively coined as "know about" and "know how" strategies, and in Hansen et al. (1999) they are respectively coined as "codification" and "personalization" strategies.

Even by the 1930s Chester Irving Barnard proposed that persons in an agency should be regarded either as *objects to be manipulated* or as *subjects to be satisfied*. This classification encompasses every management model proposed from the Taylor's pioneering scientific administration to the most recent ones. Indeed, codification management strategies bring as a corollary that the components of an agency are replaceable, as long as their knowledge is appropriately extracted, encoded and stored in a "knowledge base". Those components can therefore be deemed as objects to be manipulated just like replaceable parts of a machine. Personalization strategies, on the other hand, are founded on the principle that some components of an agency

---

[2] As discussed later, the content of a "knowledge base" is information, not knowledge. This is why we enclose it between inverted commas.

can be irreplaceable. The recursive definition of agency proposed above also suggests that the components of an agency are also agencies and therefore must have goals of their own to be achieved, hence deserving treatment as subjects to be satisfied.

Codification principles are in perfect alignment with the command-and-control agencies of Drucker (1988) and with imperative programming: agencies are assumed to run under centralized control, based on knowledge encoded in "knowledge bases". It is nevertheless the codification strategy for knowledge management that has received more attention from information technology researchers: a vast majority of the proposed computer-based tools to support knowledge management is founded on that strategy. We analyse these tools from a conceptual viewpoint, identifying their potentialities and advantages compared with the adoption of a personalization strategy, as well as their pitfalls and restrictions.

The main conceptual tools for codification-based knowledge management that we have identified are *artificial ontologies*. Since we have not identified any predominant personalization-based tool to pair up with artificial ontologies, we propose one such tool, which we believe can be the main contribution of the present book. The conceptual tool we propose is called the *structure of capability providers*, for reasons that will become clear when it is presented in Chapter 5.

Following personalization principles, the functioning of an agency is based on coordination of the activities of its components. The focus is on identifying who has the capabilities to perform specific tasks (i.e., the know-how), instead of where these capabilities have been stored and how to retrieve and interpret them.

1.2 GOALS AND MOTIVATIONS OF THIS BOOK

The book aims at providing an encompassing view of knowledge coordination. We hope that it can be read with equal interest and

1.2 GOALS AND MOTIVATIONS OF THIS BOOK

regarded as useful by information technologists as well as management professionals.

As pointed out in the introduction to Cummins and Pollock (1991), philosophical positivism has indeed left a positive legacy, namely the demand for rigour in scientific and philosophical models. Rigour and formalization are necessary (although not sufficient) conditions for clarity and precision, which in turn are necessary for testability of models and theories.

In the vast majority of cases, it is simpler to impose rigour and clarity on theories and models about *designed* entities. These theories should be essentially normative theories, or theories "about what *ought to be* the case" instead of theories "about what *is* the case" (Kyburg, 1991).

When dealing with entities whose properties, features and complexities must be investigated instead of having been designed – as is the case with people – it can be much harder to retain similar degrees of rigour and clarity to what we find in, for example, computer-based information systems and artificial intelligence systems. This can be an explanation of why so many books about knowledge management and related issues bring only general ideas and sketches of models instead of precise theories and systems that could be implemented and tested.[3] We are extremely careful not to impose on the reader the burden of complicated logico-mathematical formalization to

---

[3] In Cummins and Pollock (1991) we find a harsh attack on informality in theory building, aimed especially at philosophical theories: *Even in investigations shrouded in a façade of formalism, there is often a lamentable tendency toward hand waving when the going gets difficult. The trend is toward painting pictures rather than constructing detailed theories. Perhaps most contemporary philosophy is too vague and unfinished to satisfy even a minimal requirement of testability. The solution is not to symbolize it all in the predicate calculus. Perhaps the solution is just to strive for clarity, and not get lazy when the going gets tough. Instead of waving his or her hands, a philosopher needs to work out the details. If that cannot be done, it is an indication that the theory is wrong.*

follow our ideas, but we are also careful not to sacrifice formality because of readability. We have been bold enough not to be intimidated by the inherent difficulties of proposing a precise model for knowledge coordination involving human agents as well as software agents.

We intend our model to be used by people interested in designing systems for knowledge coordination. Hence, our model also embodies a normative view of knowledge coordination. We have tried to capture the mainstream views and desiderata for knowledge management found in highly reputable books about business practices of the matter.[4] We also pay homage to the work of Chester Irving Barnard (1936, 1938), who advanced many concepts that are being explored nowadays in knowledge management. We stress, however, that our book is about knowledge *coordination*, an issue that has rarely been dealt with explicitly before.

Our motivation to write this book was the perception that the concept of knowledge has been characterized in a rather impoverished and simplistic way by the information technology community, which does not fulfil the needs and expectations of business professionals and vice versa. We believe this must not be so and that relatively simple and effective models can be proposed for knowledge coordination that encompass a broader understanding of knowledge.

We are also interested in the reasons for the recent burst of interest in this concept in many diverse areas and the potential consequences of taking it into account in business and social agencies.

Finally, we are interested in promoting our personal view about the nature of knowledge – something to be more thoroughly explored in a forthcoming companion publication to this one.

---

[4] Some of these books, for example, are Davenport and Prusak (1998), Senge (1990) and Tiwana (1999).

1.3 INTENDED AUDIENCE

We hope that managers find this book to be a *demystifying* work about information technologies and the methods of knowledge coordination, which enables them to assess and exploit appropriately these technologies and methods.

On the other hand, we hope that information technology-oriented professionals find it clarifying and helpful in setting up their own work for the effective solution of knowledge coordination problems.

We have not avoided mathematical and logical formalization, whenever they are necessary; however, we have kept formalization to the bare minimum, just sufficient to convey our ideas with precision. The book is nevertheless self-contained and should be accessible to the interested reader, with no requirements to background technical erudition of any sort.

1.4 OVERVIEW OF CHAPTERS

In Chapter 2 we present a brief review of the main production management models and how they relate to modern knowledge management. We give special emphasis to the work of Chester Irving Barnard, who advanced all the foundational concepts regarded as important in present knowledge management models. Then we propose a working definition of *knowledge*, to be adopted throughout this book. This "working" definition differs from more general definitions such as those we find in philosophy books, for example. We have attempted to extract from a more general perspective those facets of knowledge that are relevant to agencies – hence our prior review of management models. Our working definition is therefore designed not to be a general definition, rather it is specialized and relative to the context at hand. We have been careful, nevertheless, to keep this specialized definition in accordance with a more general understanding of human knowledge that we wish to introduce. This issue will be treated fully in another book.

Our definition of knowledge depends on the concepts of agencies and agents. In Chapter 3 we propose precise definitions of these two concepts and show how these definitions support the models for knowledge coordination considered here.

In Chapter 4 we discuss *artificial ontologies*, which are the main conceptual tool for knowledge coordination based on codification.

We could not find a conceptual tool equivalent to artificial ontologies for personalization-based knowledge coordination. Since these models are indeed more akin to the needs of many agencies, in Chapter 5 we propose one such tool, coined as the *structures of capability providers*.

Finally, in Chapter 6 we present some general conclusions and prospective applications of knowledge coordination models.

REFERENCES

Barnard, C. I. (1936) *Mind in Everyday Affairs*, Cyrus Fogg Brackett Lecture, Princeton University.

Barnard, C. I. (1938) *The Functions of the Executive*, Harvard University Press.

Binney, D. (2002) "The knowledge management spectrum: The human factor," in E. Coakes, D. Willis and S. Clarke (eds), *Knowledge Management in the Sociotechnical World*, Springer-Verlag.

Cummins, R. and Pollock, J. (eds) (1991) *Philosophy and AI: Essays at the Interface*, MIT Press.

Daft, R. L. (2001) *Organization Theory and Design*, Thomson (7th edition).

Davenport, T. H. and Prusak, L. (1998) *Working Knowledge*, HBS Press.

Drucker, P. F. (1988) "The coming of the new organization," *Harvard Business Review*, **66**(1), January–February, 45–53.

Fox, M. S., Barbuceanu, M., Gruninger, M. and Lin, J. (1998) "An organizational ontology for enterprise modeling," in M. J.

REFERENCES

Prietula, K. M. Carley and L. Gasser (eds), *Simulating Organizations: Computational Models of Institutions and Groups*, AAAI/MIT Press.

Hansen, M. T., Nohria, N. and Tierney, T. (1999) "What's your strategy for managing knowledge?," *Harvard Business Review*, **106**, March–April.

Howard, D. (1996) *Discussion on Roles*, BPR List.

Kyburg, H. (1991) "Normative and descriptive ideals," in R. Cummins and J. Pollock (eds), *Philosophy and AI: Essays at the Interface*, MIT Press.

Marshall, C. (1999) *Enterprise Modeling with UML: Designing Successful Software through Business Analysis*, Addison-Wesley.

Robbins, S. P. (1990) *Organization Theory: Structure, Design and Applications*, Prentice Hall.

Scholtz, V. (2002) "Managing knowledge in a knowledge business," in E. Coakes, D. Willis and S. Clarke (eds), *Knowledge Management in the Sociotechnical World*, Springer-Verlag.

Senge, P. (1990) *The Fifth Discipline*, Currency/Doubleday.

Sowa, J. F. (2000) *Knowledge Representation – Logical, Philosophical and Computational Foundations*, Brooks/Cole.

Tiwana, A. (1999) *The Knowledge Management Toolkit: Practical Techniques for Building a Knowledge Management System*, Prentice Hall.

Verharen, E. M. (1997) *A Language-Action Perspective on the Design of Cooperative Information Agents*, The InfoLab, Tilburg University.

Watt, D. (1990) *Programming Language Concepts and Paradigms*, Prentice Hall International Series in Computer Science.

Winograd, T. (1997) "The design of interaction," in P. J. Denning and R. M. Metcalfe (eds), *Beyond Calculation: The Next Fifty Years of Computing*, Springer-Verlag.

2
Knowledge

If you don't know where you want to go, any road will take you there. – African American proverb

Knowledge must have always been a fundamental component of agencies. Early theorists, modellers and designers of agencies – especially those focusing on manufacturing industries – aimed at controlling production and hence focused their theories and models on parameters related to the quality, speed and efficiency of manufacturing. This was achieved by ignoring the value of the internal knowledge of agencies. Incentive to work was provided primarily by results outwith daily activities, such as salary increase (and the chance to acquire more of the goods that were being manufactured), assurance of comfort and wealth after retirement or quite often the prospect of abandoning those activities (e.g., by ascending to supervisory positions within the company). Perception of value was concentrated on artefacts, rather than on, for example, quality of life, regardless of ownership of goods, access to culture or autonomy as a citizen.

Recently, organizational modellers and designers have explicitly taken the notions of knowledge into account, within

---

Knowledge Coordination F. S. Corrêa da Silva and J. Agustí-Cullell
© 2003 John Wiley & Sons, Ltd ISBN: 0-470-85832-X

the context of *knowledge management*. Knowledge management has been characterized in terms of three basic activities:

- generation and acquisition;

- integration, organization and storage; and

- sharing and communication of knowledge (Fischer and Ostwald, 2001).

The procedures for knowledge management have been embodied in the following "practices" (Preece et al., 2001):

- document management systems to allow workers to find existing documents relevant to the tasks at hand;

- discussion forum systems to promote knowledge dissemination within the communities of practice and/or the communities of interest;

- capability management systems to allow managers to identify the relevant capabilities given the specific tasks to be performed; and

- "lessons-learned knowledge bases" to allow workers to retrieve previous similar cases to guide solutions to the tasks at hand.

These practices have been presented as a very loose conceptual framework, generally supported by a bunch of cases and success stories. It should be remarked that the first one is based on codification, whereas all the others are based on personalization.

Ascribing explicit value to knowledge as an asset of agencies is in line with the sociotechnical principles of organizational design, in which the fundamental resources to run an organization are considered to be human skills, knowledge and potential actions, side by side with material and technological resources.

Some important contrasts between knowledge management and production management should be taken into account:

- generally, production management is related to the end activity of a company, whereas knowledge management is related to strategic and general management. As a consequence, production management impacts directly on parameters and indices that assess the productivity of an agency, whereas knowledge management impacts on long-range subjective measures of quality and adjustment of an agency with respect to its environment (Davenport and Prusak, 1998; Senge, 1990; Terra, 2000);[1]

- knowledge is an asset of a different nature from those that are usually considered for production management. For example, the concepts of acquisition, storage, distribution and utilization of knowledge differ considerably from those of physical materials. We can run out of bolts if they are used to assemble some physical product, but we do not run out of knowing how to assemble a physical product by assembling it. Indeed, knowledge tends to improve and increase with repetition, instead of being consumed by it.

In the following sections we present a brief account of how organizational knowledge was dealt with by the first theorists of modern management – the so-called *classical school of management*. We give special emphasis to the work of Chester Irving Barnard (1886–1961), who advanced most of the concepts taken

---

[1] Exceptions to these rules are companies whose end product is knowledge, such as consultancy firms.

into account nowadays. We compare and contrast the classical school with the *socio-technical school of management* and show how the socio-technical school is more sensible to the value of knowledge than the classical school. Then we introduce a *working definition* of knowledge, which is characterized as a theoretical concept in the sense of Tuomela (1973). We criticize the most popular conceptualization of knowledge found in books and manuals of knowledge management, and then we propose a more precise definition that can be used as the foundation for knowledge coordination models.

2.1 KNOWLEDGE AND MANAGEMENT

We discuss in this section the role of knowledge in the classical and sociotechnical management models. As remarked in Chapter 1, in this we are accepting the reader's intuition and previous readings for an interpretation of the term *knowledge*. A precise personal interpretation of this term is provided in the next section, but we advise you, dear reader, to continue with the present section before reading the next one, forming your own idea of how this term should be interpreted on the way. The comparison of your interpretation with ours will be far more instructive than just reading our personal interpretation and simply accepting or bluntly rejecting it.

2.1.1 Knowledge and the classical school of management

The classical school of industrial management is usually associated with the works of Frederick W. Taylor (1911), Henry Ford, Henri Fayol (1916) and Max Weber (1947) (and their followers and contestants – Corrêa da Silva, 2001). We add to this list the works of Chester Irving Barnard (1936, 1938) for a historical appraisal of the foundations of knowledge management.

Apart from the models proposed by Barnard, we can observe

that a common feature to the classical models is how they regard and exploit the knowledge inherent to an organization. Barnard advanced the appraisal of knowledge proposed by the sociotechnical school and adopted in the majority of knowledge management books and manuals published nowadays.

2.1.1.1 Frederick W. Taylor and Henry Ford

Frederick W. Taylor was born in the United States in 1856. He worked as an engineer for steel production plants in that country from 1874 until his death in 1915.

Taylor devised an organizational model that he named *scientific management*. That model was founded on two assumptions about the behaviour of production workers (Fleury and Fleury, 1997; Fleury and Vargas, 1983; Corrêa da Silva, 2001):

- individual workers have motivations and principles – moral, ethical – that inherently conflict with the goals and motivations of the organization as a whole. For example, it is expected that workers purposefully slow down production, so that managers' expectations about their productivity decrease, hence decreasing their workload;

- individual workers act as perfectly rational economic agents aiming at optimization of their financial rewards, and therefore unnatural behaviour can be instilled in them simply by offering sufficient financial compensation.

In a brief and simplified manner, the principles of Taylor's scientific management are:

- analysis of the work to be done, aimed at finding the most efficient ways for workers to carry out their tasks;

- standardization of tools and workshop;

- worker selection based on physical and mental skills;

- supervision and control of production work by a separate group of specialized workers;

- worker appraisal and payment according to individual production.

Taylor assumed that the offer of financial advantages would be enough to get the workers to accept any interference in their work that minimized their physical effort and improved their production. His work was based on empirical analysis of the time and motions necessary to carry out production tasks. This was followed by proposed rules, tools and working methods that would optimize the mechanical effort needed to perform those tasks while allowing greater control over production, quality and speed.

Any suggestion from the workers to improve the productivity, quality or safety of production should be regarded with suspicion. Workers should be selected according to their willingness to accept and follow detailed rules of behaviour and movements. The importance of the *mental work, skills and experience* of production workers was therefore completely disregarded: the ideal worker for Taylor should behave like a deterministic, programmed machine.

Based on Taylor, production work was radically standardized. Design and decision procedures were centralized at management levels in the organization. Quality and production control were reached at very high levels, and individual workers were considered just as replaceable as any component of a machine. Production efficiency achieved very high levels, but organizations lost sensitivity to market changes, and the knowledge of

2.1 KNOWLEDGE AND MANAGEMENT 23

the workers – obtained from practical day-to-day experience – was not explored.

Henry Ford (1863–1947) was born in the United States. He was an admirable entrepreneur, obsessed with rationalization and efficiency in manufacturing (Fleury and Fleury, 1997).

Ford focused on the standardization of the production line as a whole. A factory, according to Ford, should be envisaged as a large machine in which some parts were human beings – the production line workers. Contrasting with Taylor's management principles, individual organization of work was kept as the responsibility of the workers. The main control device in Ford's factories was the conveyor belt, which organized the flow of activities and materials and allowed production control (e.g., by changing the speed of the conveyor belt).

Product design was guided by standardization and efficiency of production. The improvement in production efficiency obtained by Ford's system was outstanding: the time necessary to assemble a car was reduced from 12 hours and 8 minutes to 1 hour and 33 minutes (Fleury and Vargas, 1983). Quality control was also improved, and the cost of production was lowered, thus permitting prices to go down and products to become attainable to new markets.

It is interesting to observe that the assumptions about the behaviour and motivations of production workers were the same in Ford's and Taylor's systems. Thus, the promotion schemata and financial incentives proposed by both were similar. The results and drawbacks observed in both systems were also similar: fine quality and production control, exchangeability of production workers, increased production efficiency, all at the cost of rigidity and disregard of the value of workers' experience.

The so-called *knowledge work* was centralized at top, strategic management levels. The knowledge of production-level workers was considered valueless.

Taylor worked at the micro-level of individual workers,

whereas Ford worked at the level of the production line as a whole. Both stressed mechanization and standardization as desired features to improve control.

These two models were extremely successful at their time and became widespread. They are still in use in many agencies: the largest private bank in Brazil uses Taylor principles to organize their services – which are nowadays supported by novel technologies such as ATMs and home banking to fully automate production work, but still rely on standardization and mechanization to promote efficiency and reliability.

2.1.1.2 Henri Fayol and Max Weber

Fayol worked as an engineer from 1860 to his death in 1925 at the Commentry-Fourchambault Mining Company in France. He is acknowledged as the first management theorist. It is often cited that Fayol did for administrative activities what Taylor had done for the production activities in industrial agencies. His work was directed at building a prototypical, organizational structure in which management activities would be standardized and managers would be selected for specific activities, similar to what was proposed by Taylor for production line workers.

Fayol identified the five functions of management:

1. *planning and forecasting*;

2. *organization* of human and physical resources;

3. *command* (i.e., determining rules, tasks and guidance for workers to perform their tasks);

4. *coordination* of working efforts and initiatives; and

2.1 KNOWLEDGE AND MANAGEMENT

5. *control* (i.e., ensuring that tasks were performed according to command).

These five functions could best be achieved following the 14 principles below:

1. *division and specialization of work;*

2. clear determination of *authority and responsibility*: by "authority" Fayol meant "the right to give orders" and by "responsibility" he meant reliability in obeying orders;

3. *discipline*, which was associated with the predictability of behaviour and subordination to general organization;

4. *unity of command*: each worker should report to and receive orders from a unique manager;

5. *unity of goal*: the organization as a whole should always have a unique and clearly identified goal;

6. *subordination of individuals to the organization;*

7. *rewarding of efficiency*: performance should be measured and rewarded – not only by increased salaries and financial bonuses but also by vacations and qualitative changes in work routine;

8. *centralized command;*

9. *hierarchy-based information passing* (i.e., workers should always report only to their managers);

10. *rational lay-out*, to minimize non-productive mobility;

11. *equity* (i.e., every worker should be treated equally);

12. *stability of workforce*: productive workers should be preserved in the company;

13. *initiative*: Fayol suggests that initiative should be rewarded;

14. *team work*: managers should encourage workers to act as coordinated teams.

It is curious to find "reward to initiative" as a principle in Fayol's theory, as it seems to contradict the general idea proposed by that author. Fayol suggests the existence of a generic management model, based on his proposed functions and principles, that could be applied to any organization.

Taylor and Ford proposed the standardization of production, based on the knowledge work of management. Fayol went even further and proposed the standardization of production *and* management, based on *a priori* knowledge work embedded in the general management model proposed by him. Particular agencies should be, at their best, instances and specializations of the general model to take into account the specificities of markets, workers and products. Deviations from the proposed general model should be regarded as *organizational pathologies* to be corrected.

The works of the German sociologist Max Weber (1864–1920) relevant to management focused on the conceptualization of bureaucracy and its fundamental principles: division of work, hierarchy, rationality, standardization of rules and practices, impartiality and the standardized flow of information based on written documents (given the current technologies available in Weber's time).

Weber was interested in the social aspects of agencies. His work aimed at organizational models that would promote

2.1 KNOWLEDGE AND MANAGEMENT

social justice for workers, while ensuring efficiency in organizational activities.

Weber's model of bureaucracy had the following characteristics (adapted from Corrêa da Silva, 2001):

- *division and specialization of work*: to the point of making every task delegated to a worker liable to be reallocated to any other worker;

- *hierarchy of authority*: similar to Fayol's principles of authority, responsibility, discipline and unity of command;

- *rationality*: each worker should be assigned a position in the organization that optimized the utilization of his or her capabilities;

- *explicit rules and patterns of behaviour*: every decision and activity should be based on public and explicitly stated rules, patterns of behaviour and codes of practice;

- *documentation*: every activity within the organization should both follow documented standards and patterns and be itself documented and stored;

- *impersonality*: rules, patterns, procedures, decisions, standards and activities should be independent of the individuals belonging to the organization and related only to the structure of the organization itself.

The bureaucratic model emphasizes *control*. Like the models proposed by Taylor, Ford and Fayol, it is *mechanistic* in the sense that agencies are regarded as large machines to be designed, assembled and monitored during operation. Similar to Fayol's model, the bureaucratic model proposes a generic treatment for every organization, thus taking knowledge work away from the agencies.

Standardization is the key tool of these four models, given the assumption that the workers' knowledge should *not* be exploited. The bureaucratic model adds to this tool the requirement of *explicit documentation* of every rule, pattern, procedure, standard and activity within the organization. Hence, a successfully implemented bureaucratic model rejects value to any form of *tacit* (i.e., not linguistically expressed) knowledge.

2.1.1.3 Chester Irving Barnard

Chester Irving Barnard (1886–1961) advanced most of the concepts present in modern management theories and knowledge management (Barnard, 1936, 1938).

Barnard's work consisted of a structural and functional analysis of agencies. His basic concept was that of "formal agencies" (i.e., "conscious, deliberate and purposeful cooperation structures among people" – Barnard, 1938). By examining formal agencies, Barnard identified:

- their internal elements and variables that could generate states of equilibrium;

- the external forces that could ask for internal adjustments leading to different equilibria; and

- the functions of the executives as managers and controllers that were capable of influencing the behaviour of those elements and variables to produce desired states for the organization as a whole.

According to Barnard, individuals act on their personal motivations, goals and capabilities. When individuals have goals that they acknowledge to be beyond their personal capabilities, they

2.1 KNOWLEDGE AND MANAGEMENT

manifest a willingness to cooperate. Coordinated cooperation occurs by means of identified common purposes.

Formal agencies (e.g., whole companies, or departments within a company) have goals and motivations of their own. The capabilities of a formal organization result from the capabilities that individuals offer to share in order to achieve their personal goals. The functions of the executive must be directed to coordinate the cooperation of individuals and agencies at all levels in such way that every goal and motivation is satisfied based on the available capabilities.

This proposal of a *polymorphic hierarchy of agencies*, implicit in Barnard's work, constitutes a highly sophisticated abstract tool to analyse and to model agencies, perfectly in tune with recent information models and systems based on the concepts of *objects* and *agencies*, as discussed in the following chapters. In particular, the encompassing analyses of agencies presented in Barnard (1938) – ranging from individuals to companies to nations, states and churches – was skilfully based on this implicitly proposed hierarchy.

A simple hierarchy is a tree-like structure that characterizes the command-and-control structure of an organization. The typical representation of a simple hierarchy in an organization is an *organogram* (Figure 2.1). Higher level nodes in the organogram identify agencies that directly command and control the activity of lower level nodes linked to it. Communication flows are regulated by the organogram. Typically, the knowledge of higher level agencies has precedence over that of lower level agencies in the organogram.

A polymorphic hierarchy is a directed, acyclic graph (Figure 2.2). This is the command-and-control structure employed in many agencies nowadays (e.g., engineering companies that adopt a matrix organization, based on traditional management levels as well as project teams). Rigid hierarchies still dictate the precedence and importance of knowledge, command-and-control structures and information flow. However, more than

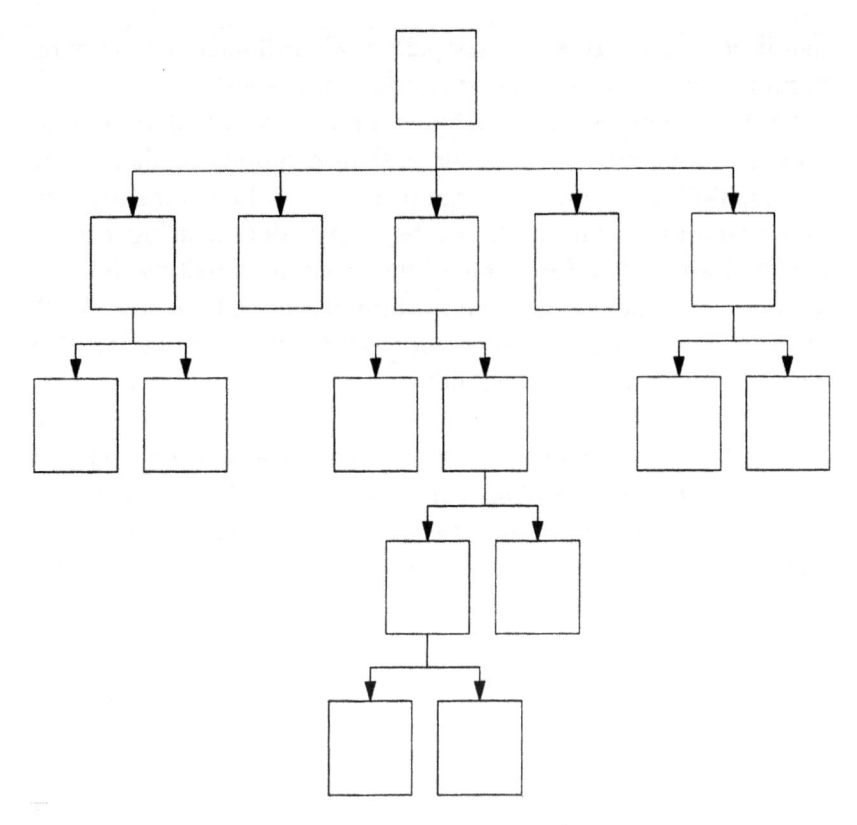

Figure 2.1 A simple hierarchy – organogram.

one hierarchy are considered in parallel, thus enriching these features in the organization.

A *network structure* is non-hierarchical (Figure 2.3). This structure is characterized as a directed graph, portraying the structures of work coordination through communication. General organizational coordination is achieved via the alignment of local and general goals, as occurs in sociotechnical agencies (Section 2.1.2).

It should become clear from the structures above that coordination results from structured communication. The three basic

2.1 KNOWLEDGE AND MANAGEMENT   31

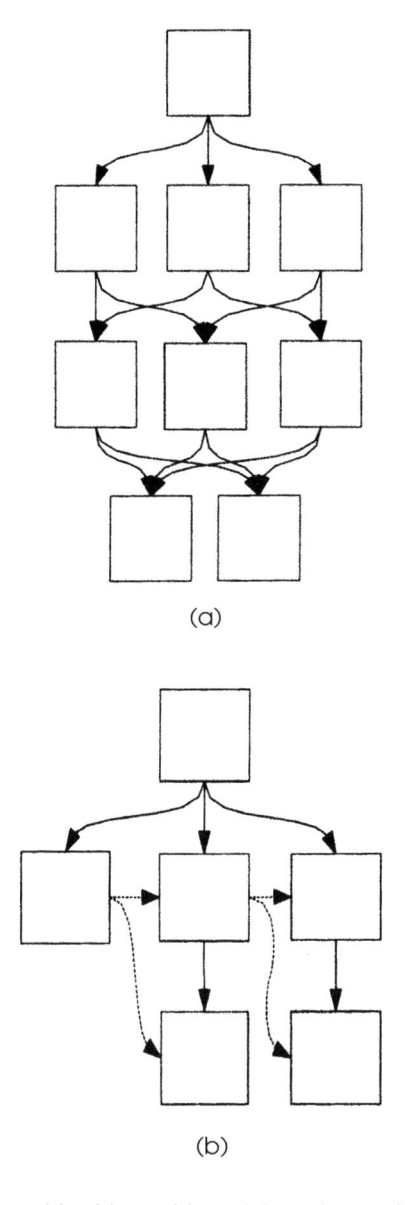

(a)

(b)

Figure 2.2 Polymorphic hierarchies: (a) polymorphic hierarchy; (b) matrix structure representing a polymorphic hierarchy: if you look only at the dashed lines or only at the full lines, you find simple hierarchies; if you consider both types of lines you have a polymorphic hierarchy.

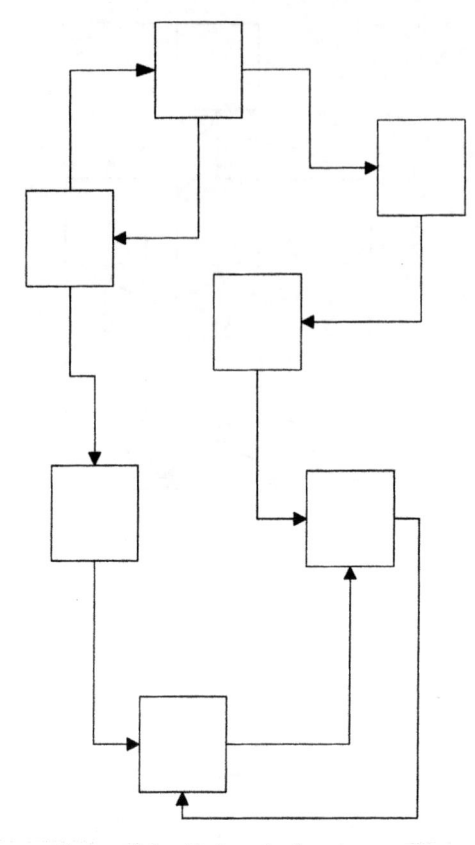

Figure 2.3 Network structure.

principles proposed by Barnard for structuring communication within formal agencies were:

- explicit and public definition of communication channels;

- hierarchical organization of information flow;

- Occam's razor – the simplest and most direct communication structures are the best.

2.1 KNOWLEDGE AND MANAGEMENT

The performance of agencies must be assessed in terms of "efficiency" and "effectiveness". These two factors are given a novel, precise meaning by Barnard:

- *effectiveness* measures the extent to which the organization is achieving its goals; and

- *efficiency* measures the extent to which the organization is avoiding undesirable side-effects in the process of achieving those goals.

An important distinction made by Barnard is on the principles used to characterize an organization. We identify two alternative principles in his work:

1. people are regarded either as *objects to be manipulated*; or

2. as *subjects to be satisfied*.

Both principles are valid as foundations for organizational models. However, as pointed out by that author, the former are more appropriate for stable situations, in which the organizational environment can be deemed static, whereas dynamic environments require adaptable and versatile agencies. For dynamic scenarios the latter prove to be more flexible and agile in their response to external changes. All models previously discussed in this chapter regard people as objects to be manipulated.

Barnard was one of the pioneers to engineer general systems of incentives to satisfy people and agencies. His "economy of incentives" proposes that objective as well as subjective incentives should be proposed to workers and agencies to bring them to act as expected:

- by *objective* incentives Barnard meant what had been taken into account up to that period (financial reward, promotion systems, improvements on working conditions, etc.);

- by *subjective* incentives he meant adjustments of the agencies to values acknowledged by society as important, so that workers would willingly contribute to an organization by understanding that the organization would be supportive of their values and principles.

We also find in Barnard (1938) an interesting digression proposing that concepts such as *organization, interaction* and *coordination* are "abstract constructs", having as their "objective" counterparts people and their actions. We place *knowledge* as another abstract construct in the sense proposed by Barnard and discuss in Section 2.2 the relevance of employing such constructs in organizational modelling. Indeed, Barnard was also a pioneer in explicitly accounting for the importance of knowledge for organizational modelling (Barnard, 1936).

Every concept, insight and method proposed by Barnard prevails in modern knowledge management.

2.1.2 Knowledge and the socio-technical school of management

The socio-technical approach to organizational design started in the 1950s at the Tavistock Institute (London), as reported in the works of F. Emery and E. Trist (cited in Coakes, 2002; Fleury and Fleury, 1997). It was largely tried and tested in the industries of Scandinavian countries.

Sociotechnical principles react to those of the classical school of management in the direction proposed by Barnard. The essence of the socio-technical school is the acknowledgement of the complexity of human agents and the consequent perception that any production system that attempts to optimize the utilization of technological resources without regard to human resources is destined to function inefficiently.

The basic tenet of the socio-technical school is that individuals must participate in decision-making and control over their local work environment. If this does not happen, then the desired levels of efficiency and effectiveness of agencies will not be achieved. The consequence of not functioning efficiently and effectively is that agencies waste their resources doing what is unnecessary and not doing what should be done.

Thus, it is the socio-technical approach to organizational design and operation that invites the explicit use of knowledge that can be found at all levels within an agency. While acknowledging that local and organizational goals must not be the same, the socio-technical school asks for a symbiotic alignment of these goals to achieve high levels of satisfaction for every agent in the organization, as well as for the organization as a whole.

Knowledge management was an implicit necessity for classical models. Employing the sociotechnical organisational model, it becomes an explicit and fundamental need for organizational design.

Cherns (1987) proposed that the socio-technical design of agencies was guided by 10 principles. These principles are discussed and analysed in Coakes (2002) and Fleury and Fleury (1997), from the viewpoint of organizational learning and knowledge management.

The goals of applying socio-technical design principles to the organization are the improvement of quality of services and products and of the organization's sensitivity and adaptability to changes in the environment. These goals are achieved by enhancing communication at all levels of the organization, rewarding continuous improvement and adaptability to environmental changes and by symbiotic alignment of local and organizational goals as mentioned before.

Cherns's principles, and how they relate to knowledge management, are presented below:

1. *Compatibility*: local and organizational goals may conflict, thus hindering the sharing of knowledge. An agency must be designed to promote the symbiotic alignment of goals, so that its components feel compelled to cooperate.

2. *Minimal critical specification*: contrasting with the classical design principles, in which every detail of working activities was determined in advance, the socio-technical school proposes the minimization of task specification. "What is done not how it is done is important" (Coakes, 2002). The flow and the dynamics of knowledge in organizations are supported by the concept of minimal critical specification. This is also the foundational principle for the information-based agencies proposed in Drucker (1988).

3. *Variance control*: an agency must respond to environmental changes effectively. This is achieved by equipping its components recursively (i.e., the agency components, the components' components, etc.) with the capability to monitor the environment and the autonomy to update their work routines accordingly. Variance control is founded on the principle that local knowledge is valuable and must be used.

4. *Boundary location*: an agency must be designed to promote knowledge sharing, rather than impeding it.

5. *Information flow*: this principle is closely related to the previous one. Agencies must be designed so that information can always efficiently reach the locations where it is needed.

6. *Power and authority*: taking the symbiotic alignment of local and organizational goals as a premise, sufficient levels of power and authority must be offered to agents, to support

their autonomy so that they can self-regulate their activities based on local as well as global knowledge.

7. *The multifunction principle*: just as agencies must adapt to environmental changes, individuals must adapt to the environment and to organizational changes.

8. *Support congruence*: we should "pay people for what they know, not what they do" (Coakes, 2002). Again, this is a very explicit statement of the value of local knowledge for an agency.

9. *Transitional organization*: agencies are essentially dynamic, and permanent change and adaptation are a natural consequence of it. Striving for the stabilization of an organizational structure is elusive and misleading.

10. *Incompletion*: as a consequence of the previous principles, one can never "complete" the design of an agency. Continuous redesign should be taken as the norm.

In his widely acclaimed book *The Fifth Discipline*, Peter Senge (1990) proposes five component technologies – which he calls "disciplines" – to understand the learning organizations (i.e., agencies that evolve autonomously in response to environmental and inner state changes).

Senge's disciplines are:

- personal mastery – methods and resources related to personal growth and development;

- mental models – explicit descriptions of concepts, models, assumptions and generalizations that determine and influence personal capabilities;

- shared vision – goals and motivations that are known to be common to more than one member of the organization;

- team learning – techniques and methods for development of individual skills by means of cooperative teamwork;

- systems thinking – structuring of agencies in terms of system dynamics, identification and reuse of patterns of system structures.

In his book, Senge introduced the fundamental concepts for understanding learning organizations. More recent books have presented techniques and case studies to suggest means of incorporating these concepts in actual management practice (Senge et al., 1994, 1999). This more practical approach can also be found in Davenport and Prusak (1998), in which case studies are interwoven with hints and general techniques to effectively incorporate knowledge management in management practice.

In Fischer and Ostwald (2001) we find some criticisms of the underlying assumptions present in these self-claimed-as practical books on knowledge management:

- Knowledge is not a commodity – the intangibility of knowledge asks for a unique methodology to acquire, generate, store, transfer, share and communicate it. We cannot acquire knowledge from an expert in the same way we can acquire a machine from a supplier, nor can we transfer knowledge from an industrial plant to another in the same way we can transport a plant's floor layout. We argue that this occurs because traditional commodities and assets considered in organizational management are observational concepts (or "objective constructs"), whereas knowledge is a theoretical concept (or an "abstract construct") (Tuomela, 1973), as developed in the following section.

- Despite the dynamics and autonomy of the learning and knowledge-based agencies, fully decentralized knowledge management is hardly the most efficient way to manage an organization.

- More than accumulating and accessing large amounts of information and knowledge, knowledge management should also be about establishing efficient ways to select, share and communicate relevant knowledge.

It is interesting to observe how these criticisms approximate the recently proposed concepts of knowledge management to the concepts and techniques proposed in Barnard's model of formal agencies. In his model we can already find: that knowledge (although not presented in this way) resides in individuals, not in databases (or knowledge bases); that one should not give up hierarchical distribution of responsibilities for information and knowledge flow; and that communication is the key issue for knowledge management.

So far, we have based our discussion on the intuitive notions of knowledge. In order to propose a precise model for knowledge coordination, however, we must be more rigorous. This is what we do next.

2.2 SO WHAT IS KNOWLEDGE AND WHERE CAN WE FIND IT?

Barnard proposed that there are two sorts of constructs used in organizational modelling:

- *abstract constructs* – such as organization, interaction, coordination and *knowledge*; and

- *objective constructs* – namely, agents and their actions.

These constructs correspond precisely to what R. Tuomela calls *theoretical* and *observational concepts*. In Tuomela (1973) we find a sophisticated argumentation to defend the indispensability of theoretical concepts in science.

According to Tuomela, observational concepts are perceptual (i.e., they can be measured or directly related to measurable events). Theoretical concepts, on the other hand, are non-perceptual: they are constructed to represent abstract entities.

Following Hempel (1958), the purpose of theoretical concepts is to establish definite connections among observable phenomena. This would lead to the "theoretician's dilemma":

1. theoretical terms either serve their purpose or they do not serve it;

2. if they do not serve it, they are dispensable;

3. if they serve their purpose, they establish relationships among observable phenomena;

4. if they establish such relationships, the same relationships can be established without theoretical terms;

5. if these same relationships are so established, theoretical terms are dispensable;

6. hence, theoretical terms are dispensable.

Step 4 is supported, for example, by Ramsey's (1931) elimination theorem, which shows that theoretical concepts can be systematically eliminated from any finitely axiomatized theory.

Following Hempel's chain of reasoning, abstract constructs should be dispensable, and any organizational model explicitly referring to them should be liable to be reconstructed in such a way as to avoid those references.

2.2 SO WHAT IS KNOWLEDGE AND WHERE CAN WE FIND IT?

Most of Tuomela's work is devoted to criticizing this step. His argumentation is based on two main points:

1. scientific theories in which theoretical terms are eliminated lose explanatory power, even though they continue to support the same results;

2. scientific theories obtained following Ramsey's procedure may be infinitely axiomatized, despite originating from finitely axiomatized ones.

We add to these points a criticism to the positivists, who state that the sole purpose of theoretical concepts is to connect observational concepts. We illustrate this criticism with two simple examples, one from mathematics and one from physics:

- *Inter-definability of propositional connectives in mathematical logics*: it is a well-known result that the logical connectives $\vee$ (disjunction) and $\wedge$ (conjunction) are inter-definable through negation ($\neg$) in mathematics:

$$A \wedge B \equiv \neg(\neg A \vee \neg B)$$

This expression can be made more lively if we attach the following meaning to its constituents:

– A: worker 1 is prepared to undertake task T;

– B: worker 2 is prepared to undertake task T;

– $\varphi \wedge \psi$: both φ and ψ are true;

– $\neg A$: worker 1 is *not* prepared to undertake task T;

– $\neg B$: worker 2 is *not* prepared to undertake task T;

- $\varphi \vee \psi$: either φ or ψ – or even both – are true;

- $\neg(\varphi)$: it is not the case that φ.

Hence, it is equivalent to say that both workers are prepared to undertake *T* and to say that it is not the case that either of them is *not* prepared to undertake *T*.

We could infer from this result that conjunction is a dispensable concept in mathematical logic, provided that we have disjunction and negation in that logic. However, this is not true in general. Indeed, it just *happens* to be true for a very specific class of logical languages, and if we concentrate on these languages solely we may confound these two fundamentally different concepts – conjunction and disjunction – due to their fortuitous inter-definability in that specific class of logics usually called *classical logic*.

Consider, for example, the following alternative meaning to be attached to the same symbols above:

- *A*: worker 1 has undertaken task *T*;

- *B*: worker 2 has undertaken task *T*;

- $\varphi \wedge \psi$: both φ and ψ are true (as before);

- $\neg A$: worker 1 is *not* free to undertake task *T*;

- $\neg B$: worker 2 is *not* free to undertake task *T*;

- $\varphi \vee \psi$: either φ or ψ – or both – are true (as before);

- $\neg(\varphi)$: it is not the case that φ (as before).

This is a sensible *non-classical* interpretation, in which the equivalence above is no longer valid: it is not the same to

2.2 SO WHAT IS KNOWLEDGE AND WHERE CAN WE FIND IT? 43

affirm that both workers are working on the same task T and that it is not the case that either is not free to undertake that task. Hence, a logical language based on this interpretation for these logical symbols would discriminate the interpretations of disjunctions and conjunctions, thus characterizing them as fundamentally different logical concepts.

In more technical terms, this non-classical logic has different interpretations for the negation symbol ($\neg$) when it occurs in front of a single statement or a longer expression. Many useful non-classical logics are built in this way; for example, *annotated logics* (Carbogim and Corrêa da Silva, 1998).

- *Interrelationship of velocity, space and time in physics*: another well-known scientific equation defines velocity in terms of space and time:

$$v(t) = ds(t)dt$$

We could conclude from this equation that velocity is a dispensable concept, since it can be replaced in the formulation of any physical model by the corresponding relation between space and time. However, as modern physics has shown, this equation holds only as an approximation of physical phenomena, valid in those situations known today as *classical physics* – it does not hold, for example, in a relativistic setting.

Similar to what happens with classical logic, it just *happens* that $v(t)$ and $ds(t)/dt$ coincide in classical physics. By no means should we conclude from this coincidence that velocity is the same as the rate of variation of space across time.

The classical management models of an organization could purport Barnard's abstract constructs as dispensable concepts. We support the view that it just happened that those concepts were definable in terms of persons and actions in those models, which is not sufficient evidence to conclude that abstract constructs are dispensable concepts. In what follows we present further arguments in favour of our view.

2.2.1 Knowledge as special information

Many texts devoted to knowledge management propose a *data–information–knowledge* hierarchy to explain the nature of knowledge (Daft, 2001; Davenport and Prusak, 1998; Tiwana, 1999).

Data are the *codification* of facts. Data start to exist when facts are *recorded* using some symbolic language. Clearly, data records can be *persistent* or *non-persistent*; that is, they can be permanently updated snapshots of reality (such as speed indication in a car velocimeter) – and hence be non-persistent – or they can record some past event for history (e.g., hieroglyphic tables indicating agricultural production in Egypt that occurred about 6,000 years ago) – and hence be persistent.

As becomes clear with the example of hieroglyphic tables above, data can outlive those who record them. Information, on the other hand, requires an agent (or agency) and is relative to agents (or agencies). *Information is data endowed with meaning*. In other words, information is interpreted data. We need an agent (or agency) to perform the interpretation.

With the fall of the ancient Egyptian civilization, the data referred to above persisted, but they lost their ability to *inform*. Those data could only achieve the status of information again after the *decodification* of the ancient Egyptian language by the Napoleonic linguists (in particular, J. F. Champollion).[2]

Interpretation is a mapping from data to their intended meaning. Information, in mathematical terms, is the domain of

---

[2] Not all of the ancient Egyptian written language has been decodified. Indeed, many suspect that all we know about ancient Egyptian writings is no more than a glimpse of the information they can convey.

2.2 SO WHAT IS KNOWLEDGE AND WHERE CAN WE FIND IT?

one such mapping. It is a subset of the universal set of data, characterized by the fact that it is mapped to some meaning.

The image of an interpretation is a set of referents that interpret some data. It is a subset of the universal set of referents, or the *ontology*[3] of an agency.

The mapping from data to meaning can be:

- $1:1$ – i.e., data and referents have a one-to-one relationship;
- $1:N$ – thus allowing a single piece of data to have more than one referent acting as its interpretation;
- $M:1$ – thus allowing a single referent to act as the interpretation of different pieces of data; or
- both $1:N$ and $M:1$ – we denote this last sort of mapping as $M:N$.

If the domain of the interpretation coincides with the universal set of data, we say we have a *total interpretation* of data, otherwise we have a *partial interpretation* of data. If the image of the interpretation coincides with the universal set of referents acknowledged by an agency, we say we have a *complete interpretation* of data, otherwise we have an *incomplete interpretation*.

Total interpretation ensures that every piece of data is interpreted and therefore deserves the status of *information*. Partial interpretation is what we have, for example, for the hieroglyphic tables above (Figure 2.4).

A complete interpretation ensures that all of the ontology of the agency is mapped into symbols. If an interpretation is incomplete then there are referents acknowledged by the agency that have no symbolic data that can express them (Figure 2.5).

---

[3] A precise meaning for *ontology* is given in Chapter 4.

Data   Meaning

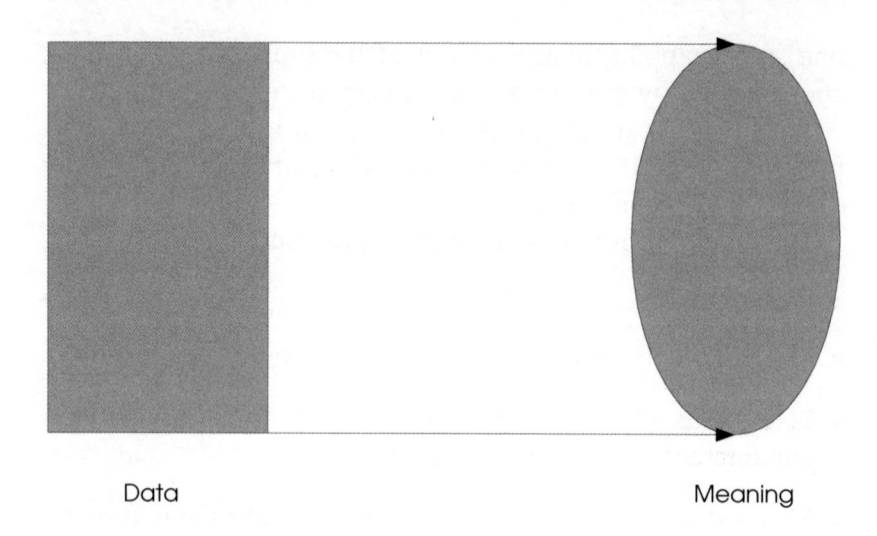

Total interpretation

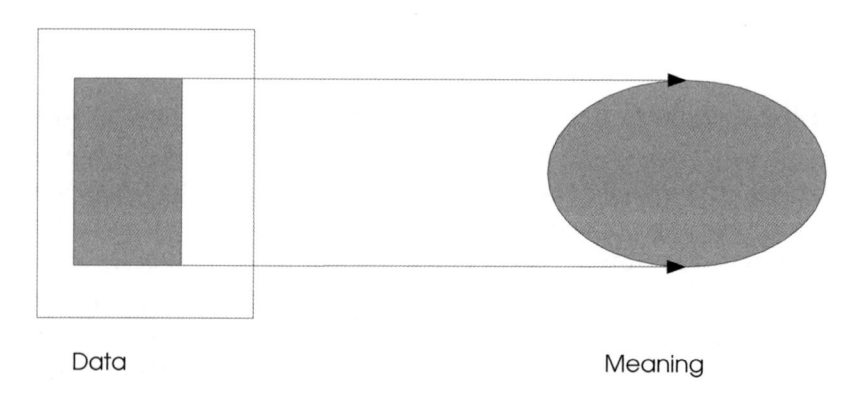

Data   Meaning

Partial interpretation

Figure 2.4 Total versus partial interpretations.

2.2 SO WHAT IS KNOWLEDGE AND WHERE CAN WE FIND IT?

Data　　　　　　　　　　　　Meaning

Complete interpretation

Data

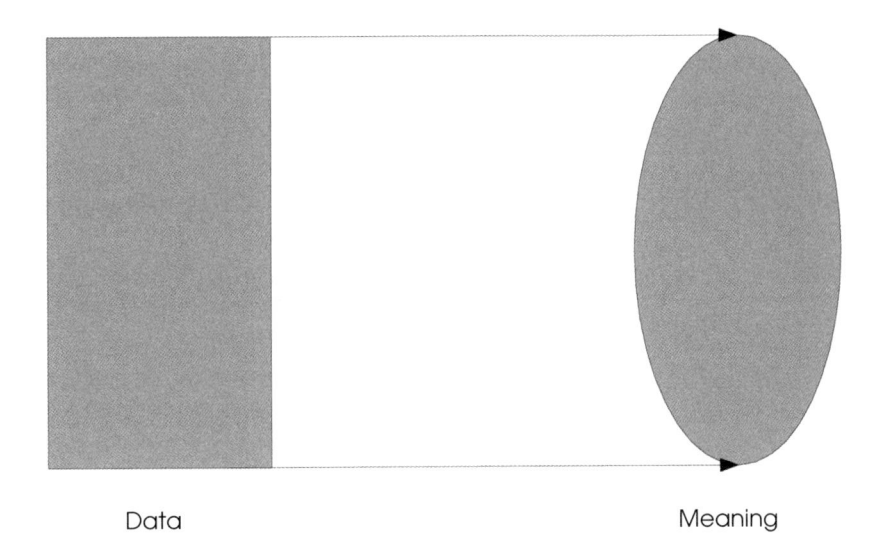

Meaning

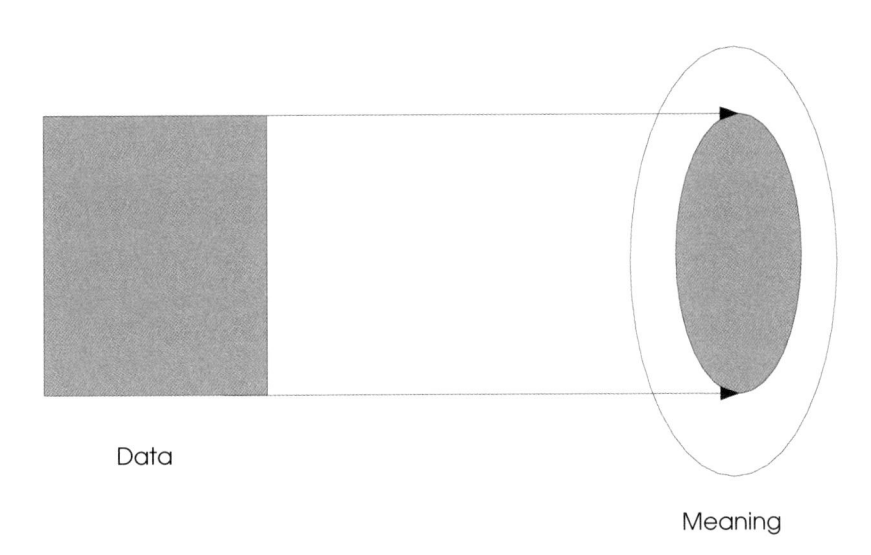

Incomplete interpretation

Figure 2.5 Complete versus incomplete interpretations.

Mappings of the form 1 : 1 are called *bijective mappings*.[4] Mappings of the form 1 : 1 are very desirable, since they do not present redundancies nor inconsistencies.

Mappings of the form 1 : N are *inconsistent interpretations*, since a single piece of data admits more than one meaning. Mappings of the form M : 1 are *redundant interpretations*, since a single referent is represented by more than one piece of data. Finally, mappings of the form M : N are both *redundant and inconsistent* (Figure 2.6).

The easiest-to-manage sort of interpreted data is *total, complete bijective information*. Unfortunately, the most common situation for human agents is the other end of the spectrum; namely *partial, incomplete, redundant and inconsistent information*. It should be the responsibility of every manager to provide incentives and the means for agencies at all levels of an organization to approximate the latter to the former type of information as much as possible.

In Tiwana (1999) we find that knowledge is *actionable information*. We interpret this as stating that knowledge is a special type of information, which endows agencies with the capability to perform actions. Knowledge is therefore relative to agencies and is what turns actions into potential capabilities for agencies.

However, this characterization of knowledge requires it to be *codifiable* – since knowledge is assumed to be information, which in turn is data. This gives no room for capabilities to act that cannot be (or have not been) documented.

In Daft (2001) and Davenport and Prusak (1998) knowledge is information constructed by an agency – by abstraction, deduction, induction or abduction. Therefore, there would be no inherent feature in knowledge, apart from how it was

---

[4] Our notion of *bijective mapping* does not fully conform with the mathematical notion of *bijective function*. A strict mathematical bijective function would be, in our terms, a *total, complete bijective mapping*.

2.2 SO WHAT IS KNOWLEDGE AND WHERE CAN WE FIND IT?

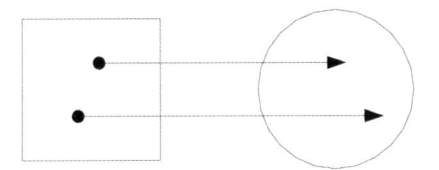

Bijective mapping

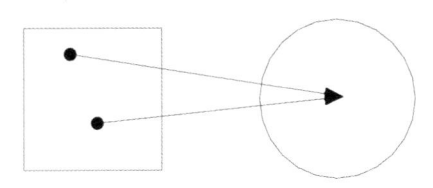

Redundant, consistent mapping

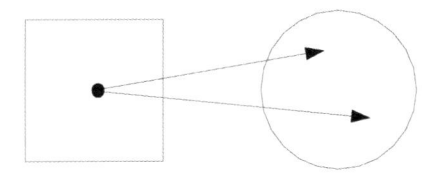

Inconsistent, non-redundant mapping

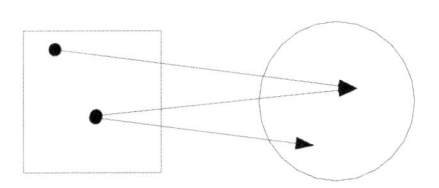

Inconsistent, redundant mapping

Figure 2.6 Mappings: bijective; redundant and consistent; inconsistent and non-redundant; and inconsistent and redundant.

constructed, to distinguish it from other sorts of information. This is quite convenient, since knowledge can thus be processed using standard data and information processing tools.

2.2.2 Knowledge as justified true belief

In Plato's *Theaetetus* we find an outstanding intellectual construction to discuss what knowledge is. At the end of that text, however, no definition of knowledge is reached. Instead, we are only told what knowledge is not. Plato ends the text with the voice of Socrates stating that knowing what knowledge is not is indeed useful, so that people can avoid believing they know something when in fact they do not.

Curiously, we find in this text a clear and carefully constructed explanation why knowledge should *not* be equated with justified true belief. Nevertheless, many recent authors have suggested that knowledge equals justified true belief, as proposed by Plato![5]

The definition of knowledge as justified true belief, despite Plato's argumentation, has been adopted as a conceptual starting point for many research projects in artificial intelligence (Corrêa da Silva et al., 1999, 2001; Delgrande and Mylopoulos, 1986). Although being an obviously incomplete definition from a philosophical standpoint, it has proven to be very convenient as a simplified account of knowledge that can be embedded in machines or, using our terminology, software agencies.

Belief is a predicate linking some information to an agency. Informally, an agency believes some piece of information if that agency conceives a possible situation in which that piece of information holds. A more formal account of belief is given in the next section.

---

[5] We can even find (e.g., Gettier, 1963) an explanation why Plato was wrong to conclude that knowledge is justified true belief.

2.2 SO WHAT IS KNOWLEDGE AND WHERE CAN WE FIND IT? 51

Defining *truth* is a complicated issue, perhaps just as complicated as defining knowledge. We consider three possible interpretations for this term:

1. Absolute truth – related to what has the property of *being*, independently of any other being who may be aware of it. This is a complex and subtle issue, which deserves a text of its own. We shall leave a thorough discussion of absolute truth for the companion publication to this one.

2. Subjective truth: truth relative to an agent. Informally, some piece of information is *true for an agent* if it holds in any conceivable situation for that agent. In this sense, truth and belief are closely related concepts.

3. Social truth: truth relative to a group of agents. Some possibilities can be considered here:

    - distributed truth – given a group of agents, some piece of information is *distributively true for the group* if, for any conceivable situation for each member of the group, there is always at least one member of the group such that that piece of information holds for that agent;

    - group truth – given a group of agents, some piece of information is *true for the group* if it is true for each member of the group;

    - common truth – given a group of agents, some piece of information is a *common truth for the group* if it is true for the group, and being true for the group is itself also information that is true for the group, and so on indefinitely.

Our homogeneous treatment of groups of agents and single agents as agencies allows us to put together subjective and social truth, which we henceforth call *agency truth*. It is not difficult to show that, for a single agent, distributed truth, group truth and common truth coincide. A formal account of agency truth is given in the next section.

We reduce "justified" to "deductively provable", thus assuming that a good justification for a belief must be a deductive proof. "Knowledge" is thus reduced to "provable agency group truth". This can be regarded as a specialization of the previously proposed definition of knowledge as information constructed by an agency.

2.2.3 Knowledge as status of an agency

Belief and agency truth are extensively analysed in Fagin et al. (1995). The formalization of these concepts is presented using multimodal logics.

Very briefly, and just to provide a taste of what it looks like,[6] we consider here the classical propositional version of what is given in Fagin et al. (1995).

Initially, we consider only the case described in Fagin et al. (1995); that is, a "flat" agency whose components are simple agents. Assume we have N agents $a_1, \ldots, a_N$, and that the information to be considered by these agents is organized in M sentences $p_1, \ldots, p_M$. Each sentence admits two possible states, namely $\top$ – indicating that it is true – and $\bot$ – indicating that it is false. Hence, for these sentences, there exist exactly 2^M possible situations to be considered, which are all possible combinations of associations of $\top$ or $\bot$ to each sentence $p_1, \ldots, p_M$. Each agent a_i regards these 2^M alternatives precisely as possible situations.

A logical expression involving the sentences p_j selects a subset of the set of possible situations. For example, the logical

---

[6] The interested reader is strongly advised to consult Fagin et al. (1995) for further details. All definitions given here are directly extracted from that reference.

2.2 SO WHAT IS KNOWLEDGE AND WHERE CAN WE FIND IT?

expression $(p_1 \wedge \neg(p_2))$ selects those possible situations in which the value $\top$ is assigned to p_1 and the value $\bot$ is assigned to p_2.

To each agent is assigned a directed graph that connects states, so that from a given state that agent can "see" those states with a direct path to them. An agent a_i is defined to *believe* in a sentence p_j at a given state $s_r \in \{s_1, \ldots, s_R\}$, $R = 2^M$, if the value $\top$ is assigned to p_j in at least one possible situation that the agent can "see" from s_r.

We say that a state s_a is *reachable* from a state s_b for an agent a_k if the agent can "see" s_a from s_b (i.e., if there is a connection from s_b to s_a).

This notion of belief can be extended to logical sentences. An agent a_i is defined to believe in a logical expression φ at a given state s_r if the value $\top$ is assigned to the expression in at least one possible situation that the agent can "see" from s_r (Figure 2.7). If the usual extension of assignment of values $\top$ and $\bot$ to sentences is assumed (e.g., if $\varphi = (p_1 \wedge p_2)$), then the value $\top$ is assigned to it only in those states in which it is also assigned to both p_1 and p_2.

We assume here a classical language as the underlying logical language to model the world. The meaning of this is that in any state s_r one and only one value belonging to $\{\top, \bot\}$ is assigned to each sentence p_j. We also assume a classical interpretation for the negation symbol ($\neg$): if the value $\top$ is assigned to φ, then the value $\bot$ is assigned to $\neg \varphi$, and vice versa. With these two assumptions, knowledge is easily defined based on the concept of belief. An agent a_i is defined to *know* a logical expression φ at a given state s_r if it cannot believe in $\neg \varphi$ at that same state. In other words, the value $\top$ is not assigned to the expression $\neg \varphi$ in any possible situation that the agent can "see" from s_r. Since the language is classical, as well as the interpretation of the negation symbol, this entails that the value $\top$ is assigned to the expression φ in *every* possible situation that the agent can "see" from s_r (Figure 2.8).

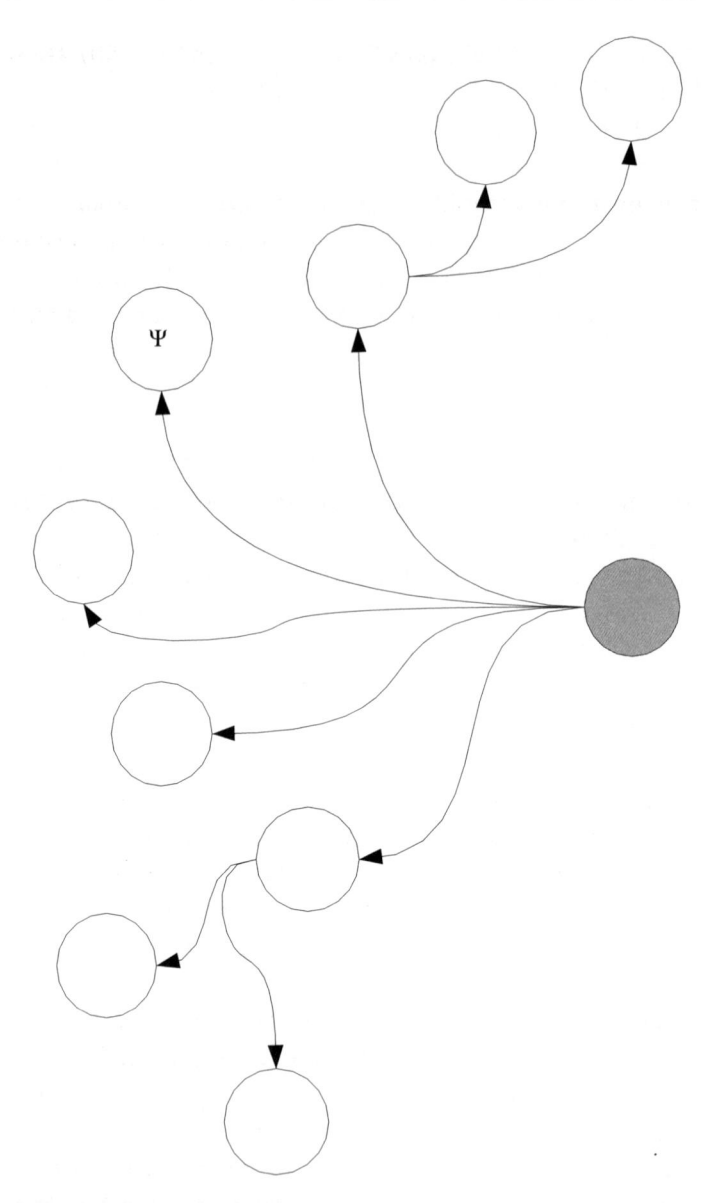

Figure 2.7 Belief of an agent. Circles represent states and arrows represent reachability relations. The tinted circle represents the present state s_r. An expression Ψ inside a circle indicates that the value $\top$ is assigned to that logical expression in the corresponding state.

2.2 SO WHAT IS KNOWLEDGE AND WHERE CAN WE FIND IT?

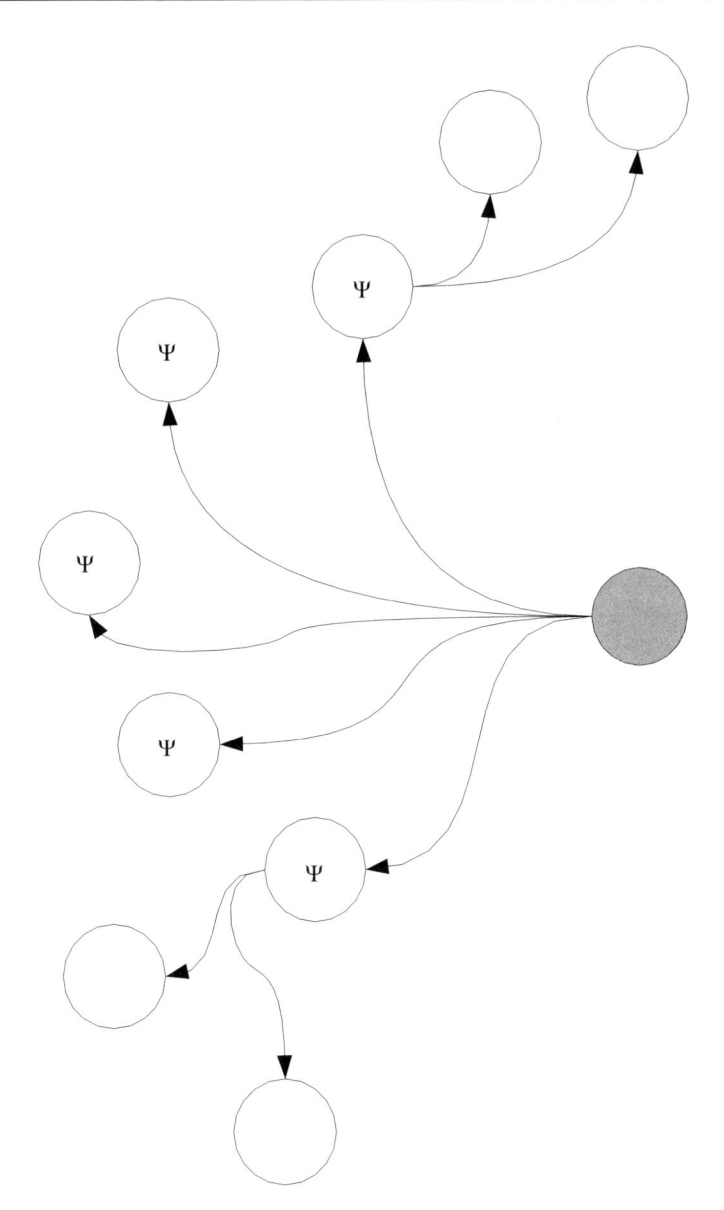

Figure 2.8 Knowledge of an agent. Circles represent states and arrows represent reachability relations. The tinted circle represents the present state s_r. An expression Ψ inside a circle indicates that the value $\top$ is assigned to that logical expression in the corresponding state.

If the language – or the assumed interpretation of negation or some other symbol in the language – is not classical, quite often this last statement is assumed as a formal definition of knowledge (i.e., having the value $\top$ assigned to it in every reachable possible situation). This is a convenient assumption from the formal point of view; however, it seems to have an intuitive motivation only for classical languages.

Given a set of agents, we define *distributed belief* as follows: the whole set distributively believes in φ at a given state s_r if $\top$ is assigned to φ in at least one state reachable from s_r by at least one agent in the set of agents.

Similarly, we define *distributed knowledge*: given a set of agents and a state s_r, we have the set of states reachable by all agents from s_r. This is given by the intersection of the sets of states reachable by each agent from s_r. The set of agents distributively knows φ at s_r if $\top$ is assigned to φ in every state reachable by all agents.

Group belief and *group knowledge* are easier to describe: agents believe in φ as a group at s_r if each agent independently believes in φ at s_r. A set of agents knows φ as a group at s_r if each agent independently knows φ at s_r.

Finally, we take into account that knowing something is also information, and therefore an agent may know that it knows something, or that some other agent knows something, etc.

We define *common knowledge* as follows: agents have common knowledge about φ at s_r if they know φ at s_r as a group, and they know that they know φ at s_r as a group, and they know that they know that they know φ at s_r as a group, and so on.

Intuitively, group knowledge is broadcast information and common knowledge is publicly broadcast information. If a piece of information is given to each worker in a company by telephone and independently, it becomes group knowledge. On the other hand, if a piece of information is announced on the notice board in front of the office, it becomes common knowledge.

Depending on the application, agency knowledge can be defined as group knowledge or common knowledge. If it is defined as group knowledge, then it is intuitive to generalize this definition to "non-flat" agencies. Given any agency, the knowledge of the agency as a whole becomes simply the group knowledge of the agents that compose it, either directly or by composing smaller agencies that compose the agency.

The topology of the graph that connects states determines the properties of the knowledge and belief of each agent. For example, if the graph is reflexive (i.e., every state is reachable from itself, therefore the agent can always "see" the present state), then whatever the agent knows holds for the situation in which the agent is. If the graph is reflexive, symmetric and transitive, then the possible situations are organized in equivalence classes, a mathematical property that permits the derivation of interesting features of an agent's knowledge.

It is interesting to contrast this definition of knowledge with the characterization proposed in Tiwana (1999). Here, knowledge is a special type of connection between agents and pieces of information: an agent can *know about* some countries in the Middle East that can influence the price of petroleum, or the weather that is usually warmer in summer than in winter, etc. The notion of knowledge proposed in Tiwana (1999) is more akin to organized capabilities: an agent can *know how* to ride a bicycle, write in Chinese, analyse the status of economic markets from the fluctuation of stocks, etc.

2.2.4 Knowledge as the skill to provide meaning to data

Peter Drucker (1988) suggests that *knowledge* is what transforms data into information. According to this view, learning could be understood as the process of transforming partial, incomplete, redundant and inconsistent information into total, complete and bijective information.

This view does not seem to be prevalent one. Not even Drucker attempts to explore it very much in his writings. It is an interesting proposal, nevertheless, that departs from the previous one by associating knowledge with the metalanguage that characterizes data and information. Knowledge is not information and does not need to be codifiable, although it *can be* codifiable in some special cases.

2.2.5 Knowledge as the capability to change the world

The knowledge as in *know how*, rather than as in *know about*, is directly related to *action*. Knowledge in this sense is the capability of an agency to change the world. As proposed in Liebowitz (2001), knowledge is "the capability to act".

We avoid the use of the term *information* to characterize knowledge, in order to encompass non-codifiable knowledge in our characterization.

To change the world means to update the value assigned to some sentence that describes it. A sentence can be related to some physical property of the world, some abstract property, the knowledge of an agency, some belief of an agent, the capability of an agency, the goals and motivations of an agency, etc. When an agency uses its knowledge, something changes in the world. However, an agency is not obliged to use all its knowledge all the time. The actions of an agency are restricted to its capabilities – and hence its knowledge – and occur based on the motivations, goals and plans of the agency.

Knowing about is characterized by the status of an agency, as proposed above. *Knowing about* is itself the result of an action (that can be the deduction of some logical expression using some inference rules. An inference rule is therefore a codifiable piece of knowledge in our proposed sense of *knowing how*; namely, it describes how an agent *can* generate new logical expressions from existing ones.) Hence, *knowing about* is the result of *knowing how*.

2.2 SO WHAT IS KNOWLEDGE AND WHERE CAN WE FIND IT?

We list below some concrete cases of knowledge as capabilities, for human agents as well as for software agents:

- Human agents:

  - *Craftsmanship* – a luthier takes some pieces of wood and some tools and makes a violin. This is a capability to change physical properties of the world. The luthier knows how to make violins.

  - *Manufacturing* – Yamaha makes musical instruments (e.g., the Clavinova series of digital pianos). Differently from above, it is Yamaha as an agency that holds the capability of making Clavinova pianos. It is Yamaha – the company – that knows how to make Clavinova pianos.

  - *Financial auditing* – a financial auditor knows how to assess a company. The auditor takes a company and prepares a report. The change in the physical world resulting from preparing the report is irrelevant. The auditor produces information out of the analysed company.

  - *Learning* – an agency updates its set of capabilities by means of learning and it takes specialized *learning skills* for effective learning. Knowing to learn is knowing to change our knowledge.

- Software agents:

  - *Data processing* – conventional programs produce new data from previously existing data. We can say that a program (e.g., a compiler for some programming language, or an application to calculate the annual balance for a company given its financial results and reports) embeds the knowledge to produce those data.

- *Expert systems* – the boundary that separates expert systems and conventional programs is hard to identify. Expert systems are programs that explicitly embed the capabilities of a human expert to execute a certain task. Given a set of data, represented in an appropriate form, an expert system performs actions in a similar way to those the human expert would perform, with comparable degrees of efficiency.

- *Machine learning and meta-programming* – some programs can adapt their behaviour dynamically according to the data that are given to them. The technology to build such programs is generically called machine learning. Among the most well-known learning programs are those based on *neural networks*, whose architecture is inspired in the organization of biological neural systems. Other programs, instead of processing data directly, generate specialized programs to do so. A program that produces another program given a set of data is called a meta-program. Machine learning and meta-programming require the knowledge to produce or to adapt programs given a set of data.

2.2.6 Knowledge and agencies

The knowledge of an agency is usually more than just the knowledge of its components. The difference results from knowledge sharing and cooperation among the components of an agency. Knowledge coordination is how knowledge sharing is organized.

Considering knowledge as the capability to change the world, as proposed above, the composition of capabilities from different agencies may generate novel capabilities. As a very simple example, consider the task of moving a grand piano from one room to another (assume the piano has no wheels). This may be beyond the capabilities of any single person, but the organized

2.2 SO WHAT IS KNOWLEDGE AND WHERE CAN WE FIND IT?

composition of the capabilities of a few well-built individuals makes a new capability appear.

Composition of capabilities can occur based on two main principles, as discussed in Corrêa da Silva et al. (1999, 2001) for software agencies. Common to these two principles is the idea that, at a given time, we can always identify an agency that requests a service and an agency that provides a service, which we call, respectively, a capability client and a capability server:

1. The capability server can act as an oracle. In this case, the capability client requests a service and provides the necessary input (which in Corrêa da Silva et al. (1999, 2001) takes the form of computational data structures), and the capability server performs all the jobs to carry out the request. Relevant issues in this case are:

   - How to describe what capabilities may be required by a client, and how to match these capabilities with those advertised by potential servers. This is discussed in Chapter 5.

   - What agencies to consult to act as capability servers, given a certain task to be executed by a capability client. This selection can be done based on *structures of capability providers*, which are also discussed in Chapter 5.

   - How to communicate what service must be done and what is available to execute that service, and how to interpret and use the output of a server. This requires that client and server communicate using a common language, based on the same ontology. This is discussed in Chapter 4.

   Having the capability server act as an oracle makes it indispensable in a structure of service offerings. This is our reconstruction of what is called the personalization strategy for knowledge sharing in Hansen et al. (1999), the know-how approach for knowledge management in Daft (2001) and formal organizations that treat agencies as subjects to be satisfied in Barnard (1938).

As will be seen in Chapter 5, structures of capability providers formalize the idea of communities of practice. A community of practice is an agency whose components informally cooperate for mutual problem-solving. Informality, spontaneity and self-organization are considered key aspects of communities of practice that enable and foster the free flow of exchange of information and help within a community. These features quite frequently have been misunderstood as stating that communities of practice cannot be fostered, managed or controlled. Structures of capability providers constitute an effective tool to manage communities of practice.

Structures of capability providers can also be employed for automated sharing of capabilities among software agencies. Indeed, this was the original application conceived for this tool (Robertson et al., 2000). They can therefore be employed with success in agencies whose components are human as well as software agencies.

2. The capability server can act as a surrogate "knowledge base" for the capability client. In this case, it is the client that executes the task, but its "knowledge base" is extended by what is contained in the server. Notice that the meaning of knowledge in this case becomes closer to information than in the previous case. Indeed, it is the information contained in the server that extends the client's own set of capabilities. Relevant issues in this case are:

- How to identify what information is missing for the client to execute a task.

2.2 SO WHAT IS KNOWLEDGE AND WHERE CAN WE FIND IT?

- How to find the missing information inside the "knowledge base" of potential servers.

- How to retrieve information from the servers, and how to interpret this information.

- How to select what information to use.

The last issue relates to strategies for problem-solving that must be embedded in the client agency. The other three issues require that client and server use a common language to encode the information on which they have the capability to act, as well as the result of employing their capabilities, and that this language is based on the same ontology. This is thoroughly discussed in Chapter 4.

If the information contained in the "knowledge base" of a potential client is exported to a generic "knowledge base", the original server can become dispensable in this case. This is our reconstruction of what is called the codification strategy for knowledge sharing in Hansen et al. (1999), the *know about* approach for knowledge management in Daft (2001) and formal organizations that treat agencies as objects to be manipulated in Barnard (1938).

Ontologies and knowledge coordination based on codification are discussed in Chapter 4. Structures of capability providers and knowledge coordination based on personalization are discussed in Chapter 5. Our concept of knowledge as the capability to change the world relies on the concepts of agent and agency, as well as these two strategies for knowledge coordination. Before we continue, we must discuss more carefully what is meant by an agent and an agency. This is what is presented in the next chapter.

REFERENCES

Barnard, C. I. (1936) *Mind in Everyday Affairs*, Cyrus Fogg Brackett Lecture, Princeton University.

Barnard, C. I. (1938) *The Functions of the Executive*, Harvard University Press.

Carbogim, D. V. and Corrêa da Silva, F. S. (1998) "Annotated logic applications for imperfect information," *Applied Intelligence*, **9**, 163–172.

Cherns, A. (1987) "Principles of sociotechnical design revisited," *Human Relations*, **40**(3), 153–162.

Coakes, E. (2002) "Knowledge management: A sociotechnical perspective," in E. Coakes, D. Willis and S. Clarke (eds), *Knowledge Management in the Sociotechnical World*, Springer-Verlag.

Corrêa da Silva, F. S., Vasconcelos, W. W., Agustí-Cullell, J., Robertson, D. S. and Melo, A. C. V. (1999) "Why ontologies are not enough for knowledge sharing," in *12th International Conference on Industrial and Engineering Applications of Artificial Intelligence and Expert Systems, Egypt*.

Corrêa da Silva, F. S., Vasconcelos, W. W., Robertson, D. S., Brilhante, V., Melo, A. C. V., Finger, M. and Agustí-Cullell, J. (2001) "On the insufficiency of ontologies: Problems in knowledge sharing and alternative solutions," *Knowledge Based Systems*, **14**(7).

Daft, R. L. (2001) *Organization Theory and Design*, Thomson (7th edition).

Delgrande, J. P. and Mylopoulos, J. (1986) "Knowledge representation: Features of knowledge," in W. Bibel and P. Jorrand (eds), *Fundamentals of Artificial Intelligence: An Advanced Course*, Springer-Verlag (LNCS 232).

Davenport, T. H. and Prusak, L. (1998) *Working Knowledege*, HBS Press.

Drucker, P. F. (1988) "The coming of the new organisation," *Harvard Business Review*, **66**(1), January–February, 45–53.

Fagin, R., Halpern, J. Y., Moses, Y. and Vardi, M. Y. (1995) *Reasoning about Knowledge*, MIT Press.

Fayol, H. (1916) "Administration industrielle et générale," *Bulletin de la Société de l'Industrie Minerale* (Dunod) [in French].

REFERENCES

Fischer, G. and Ostwald, J. (2001) "Knowledge management: Problems, promises, realities, and challenges", *IEEE Intelligent Systems*, 60–72.

Fleury, A. and Fleury, M. T. L. (1997) *Aprendizagem e Inovação Organizacional*, Atlas [in Portuguese].

Fleury, A. and Vargas, N. (eds) (1983) *Organização do Trabalho*, Atlas [in Portuguese].

Gettier, E. L. (1963) "Is justified true belief knowledge?", *Analysis*, **23**(6), 121–123.

Hansen, M. T., Nohria, N. and Tierney, T. (1999) "What's your strategy for managing knowledge?", *Harvard Business Review*, **106**, March–April.

Hempel, C. G. (1958) "The theoretician's dilemma," in H. Feigl, M. Scriven and G. Maxwell (eds), *Minnesota Studies in the Philosophy of Science*, University of Minnesota Press (Vol. II).

Liebowitz, J. (2001) *Knowledge Management: Learning from Knowledge Engineering*, CRC Press.

Plato, "Theaetetus," in E. Hamilton and H. Cairns (eds), *The Collected Dialogues*, Princeton University Press (16th printing).

Preece, A., Flett, A., Sleeman, D., Curry, D., Meany, N. and Perry, P. (2001) "Better knowledge management through knowledge engineering," *IEEE Intelligent Systems*, 36–42.

Ramsey, F. (1931) "Theories," in *The Foundations of Mathematics and Other Logical Essays*, Littlefield, Adams and Co.

Robertson, D. S., Agustí-Cullell, J., Corrêa da Silva, F. S., Vasconcelos, W. W. and Melo, A. C. V. (2000) "A lightweight capability communication mechanism," *13th International Conference on Industrial and Engineering Applications of Artificial Intelligence and Expert Systems, New Orleans*.

Senge, P. (1990) *The Fifth Discipline*, Currency/Doubleday.

Senge, P., Kleiner, A., Roberts, Ch., Ross, R. and Smith, B. (1994) *The Fifth Discipline Field Book: Strategies and Tools for Building a Learning Organization*, Currency/Doubleday.

Senge, P., Kleiner, A., Roberts, Ch., Ross, R., Roth, G. and Smith, B. (1999) *The Dance of Change: The Challenges of Sustaining Momentum in Learning Organizations*, Currency/Doubleday.

Silva, R. O. (2001) *Teorias da Administração*, Thomson [in Portuguese].

Taylor, F. W. (1911) *Principles of Scientific Management*, Harper & Row.
Terra, J. C. C. (2000) *Gestão de Conhecimento: O Grande Desafio Empresarial*, Negocio [in Portuguese].
Tiwana, A. (1999) *The Knowledge Management Toolkit: Practical Techniques for Building a Knowledge Management System*, Prentice Hall.
Tuomela, R. (1973) *Theoretical Concepts*, Springer-Verlag.
Weber, M. (1947) *The Theory of Social and Economic Organisation*, Free Press.

3
Agents

n
Othi
n

g can

s
urPas
s

the m

y
SteR
y

of

s
tilLnes
s – e.e. cummings

Knowledge Coordination   F. S. Corrêa da Silva and J. Agustí-Cullell
© 2003 John Wiley & Sons, Ltd   ISBN: 0-470-85832-X

As suggested in the previous chapters, knowledge coordination must be founded on the concepts of *knowledge* and *action*, and both these concepts are founded on the concept of *agency*.

In the present chapter, we detail the conceptualization of agency necessary to discuss knowledge coordination. We must be able to represent concepts such as *information*, *knowledge*, *organizational modelling* and *flow of knowledge and information*.

The term *agent* – which we generalize here as *agency* – has proven to be very flexible. It has been used successfully to comprise important concepts in many areas, such as economics, sociology, psychology, and software systems design and implementation (Fisher et al., 1997; d'Inverno et al., 1997; d'Inverno and Luck, 2001; Luck and d'Inverno, 2001; Muller, 1998; Russell and Norvig, 1995). It has also been used as a foundational concept for organizational modelling and design (Bernus and Nemes, 1999; Carley, 1995; Prietula et al., 1998).

Initially, we present a personal view about agents. This view is strongly influenced by d'Inverno and Luck (2001) and Luck and d'Inverno (2001), but we adapt their characterization of agents – built primarily to deal with software agents – to admit all the sorts of agents that can occur in an organization. Then, we show how these agents can be used to constitute agencies. Finally, we propose an agency-based general definition of management, which will be the basis for our discussion about knowledge coordination.

3.1 AGENTS FOR KNOWLEDGE MODELLING

Agency will be the central concept in our language. Our goal is the coordination of knowledge among and within agencies. Agencies can be software tools, robots, individuals, whole departments within a company, whole companies within a market, nations, etc.

3.1 AGENTS FOR KNOWLEDGE MODELLING

We build our conceptualization of agency from the bottom up. In the following paragraphs we first build some foundational concepts, then we assemble these concepts to build progressively more sophisticated entities, until we reach the concept of *autonomous agencies*.

The basic concept we use is that of an *attribute*. An attribute is the name of a property, feature or predicate. Each attribute has an associated set of *values*, called *instances* of the attribute. For example, *colour* can be an attribute and *green* an instance of *colour*.

Attributes are represented as *typed unary predicates*. The *type* of an attribute characterizes the set of *values* it admits. In the example above, we can represent the attribute *colour* with the predicate p. Assuming that the colours for this attribute are limited to *red, green* and *blue*, the *type* of p is given by the set {*red, green, blue*}. The expression $p(X)$ is a representation of the uninstantiated attribute p, where X is a typed variable, whose values can be *red, green* or *blue*. The expression $p(green)$ is a representation of an instance of $p(X)$.

Attributes are independent of each other. If necessary, we can easily build dependencies among attributes, represented as *clausal theories*, such that the instantiation of some attributes determines the instantiation of others.

Clausal theories are logical theories suitable for automated theorem proving. A clausal theory must be written in a specific logical language and have companion proof procedures that can be implemented as a computer program. Clausal theories together with computational proof procedures form the scientific discipline called *logic programming*.

Logic programming is founded on the observation that the proof of a logical theorem with no appeal to mathematical genius and creativity is an intellectual process similar to step-by-step execution of an imperative program. The initial theory works as a set of axioms, and the proof procedures generate new logical expressions until some goal expression (a theorem) is obtained, or the impossibility to generate the goal expression is established.

Expressions can be *specialized*. The classical example to present the concept of specialization is the deduction that Socrates is mortal, based on the assumptions that Socrates is a man and that every man is mortal. "Every man is mortal" is a general expression. A specialization of this expression is the sentence "if Socrates is a man, then Socrates is mortal". This sentence can be matched with the fact "Socrates is a man", thus permitting the conclusion that "Socrates is mortal".

Specialization is a mechanism for parameter passing among sentences. If the sentence "every man is mortal" is written as "if X is a man then X is mortal", in which X is a variable, it becomes easier to realize that "Socrates is a man" is the specialization of that sentence in which the value "Socrates" was ascribed to X.

Logic programming is an implementation of computations that coincide with the proof of theorems. In logic programming, the computation steps that occur during the execution of a program admit double interpretation – as execution steps of a program and as proof steps of a theorem. Formal specification of problems as logic programs are themselves executable programs that solve these problems.

Our brief presentation of logic programming is based mainly on the excellent (and marvellously concise) text by Krzystof R. Apt (1994). This is not a recent text, but it is still a fundamental reference for the theoretical foundations of logic programming.

The birth of logic programming is usually marked by the article by J. A. Robinson (1965), in which we find the theoretical results that gave room to the construction of programming languages based on specialization of sentences and proof of theorems as outlined above. During the first half of the 1970s the research groups led by A. Colmerauer in France and R. Kowalsky in the UK built the first implementations of the programming language Prolog, which is founded on the results presented in Robinson (1965).

3.1 AGENTS FOR KNOWLEDGE MODELLING

Logic programming gained popularity when it was adopted as the basic programming paradigm for the Japanese Fifth Generation Project during the 1980s. During that period, this programming paradigm – and the programming languages built based on it – achieved theoretical as well as engineering maturity. The theory of logic programming matured with the characterization of classes of programs with well-founded formal semantics and with the proof of many results characterizing their expressive power. The engineering of logic programming matured with the implementation of a plethora of efficient logic programming languages, many of which were freely available through open source licenses.

Some preliminary concepts must be established for a technical introduction to logic programming. We make here a concise and simplified presentation of these concepts, to avoid distraction from our main discussion. We recommend the interested reader to consult some textbooks on mathematical logic to complement our brief presentation (e.g., the excellent books by E. Mendelson, 1987 and J. Shoenfield, 1967).

We start with the concept of first-order language. A first-order language is formed by the following elements:

- *variables* – a countable set of untyped variables;

- *n-ary functions* – n is a finite natural value ($0 \leq n < \infty$); if $n = 0$ then the function denotes a constant value; if $n > 0$ then the function denotes a mathematical n-ary function: given the values of the n arguments, the value of the function is determined;

- *n-ary relations* – n is a finite natural value as above; if $n = 0$ then the relation denotes a logical proposition, which can be either *true* or *false* (e.g., the sentence "Mary is pregnant"); if $n > 0$ then the relation denotes a set; if $n = 1$ then it is a set of "simple" elements; if $n = 2$ then it is a set of ordered pairs of "simple" elements, etc.;

- *propositional constants* – $\bot$ to represent *false* and $\top$ to represent *true*;

- *logical connectives* – we use here the connectives $\leftarrow$ (logical implication) and $\wedge$ (conjunction); logic programming can also be extended to use negation ($\neg$), as presented in Apt (1994), but we are not using this connective in our presentation.

These elements are used to build *terms* and *clauses*. A term is an expression formed by:

- a variable X;

- an n-ary function followed by n terms; if $n = 0$ then the term is the corresponding constant (e.g., c); if $n > 0$ then the term denotes the application of a function to a list of terms. For example, if f is a ternary function we can build the term $f(c, X, f(Y, b, X))$, in which X and Y are variables and b and c are constants.

A *positive basic literal* is an n-ary relation followed by n terms. For example, if p is a binary relation, then we can build the positive basic literal $p(f(b, c, X), f(Y, c, c))$.

In this text we only consider *Horn clauses*. A Horn clause is an expression of one of the following forms:

1. $\bot \leftarrow \top \wedge \Psi$, in which Ψ is the conjunction of m positive basic literals ($0 \leq m < \infty$); these clauses are named *Horn queries*;

2. $\varphi \leftarrow \top \wedge \Psi$, in which Ψ is as above and φ is *one* positive basic literal; these clauses are named *program clauses*.

3.1 AGENTS FOR KNOWLEDGE MODELLING

A *logic program* is a finite non-empty set of program clauses.

With this in hand, we can now provide a better explanation of the double interpretation of clauses. Let us consider the program clause:

$$\varphi_0 \leftarrow \top \wedge \varphi_1 \wedge \cdots \wedge \varphi_m$$

We can interpret this clause as expressing that, if $\varphi_1, \ldots, \varphi_m$ are true, then so is φ_0. This is the *logical interpretation* of the clause. We can also interpret this clause as stating that the expression φ_0 is solved by the execution of the program steps $\varphi_1, \ldots, \varphi_m$. This is the *operational interpretation* of the same clause.

Since there is a formal equivalence between these two interpretations, we can switch between interpretations according to our will.

As suggested above, parameter passing in logic programming is based on specialization of variables within clauses. It is assumed that all variables in every clause are *universally quantified*. Intuitively, this means that the clauses express valid relations among all values that can be assigned to each variable. For example, the clause:

$$p(X) \leftarrow \top \wedge q(X)$$

states that for every value that can be assigned to X, if the value belongs to the relation q then it also belongs to the relation p. In other words, the set of values characterized by q is a subset of the set of values characterized by p.

Each specific value of X that is proved to belong to q is also an element of p. This is the essence of logic programming: elements of relations are determined by the specialization of program clauses. Usually, program clauses are far more complex than the one presented above, thus specialization can entail highly sophisticated patterns of parameter passing.

Assuming that the constant *a* belongs to *q*, the specialization of the clause $p(X) \leftarrow \top \wedge q(X)$ is performed by the *substitution* of the occurrence of *X* by *a*, producing the "new" clause:

$$p(a) \leftarrow \top \wedge q(a)$$

This clause has no variables. It is a very specialized instance of the previous clause, stating that if *a* belongs to *q* then it also belongs to *p*.

If we replace *q* by "man", *p* by "mortal" and *a* by "Socrates", we have the same example presented at the beginning of this section.

Formally, a substitution is a finite set of pairs (X_i, t_i), in which each X_i is a variable and each t_i is a term. A substitution is well formed if each variable X_i occurs in only one pair and if $X_i \neq t_i$ in every pair. In the remainder of this section let us denote substitutions by Greek letters θ, α, β, ... For example, the substitution:

$$\theta = \{(X, a), (Y, X)\}$$

is a well-formed substitution.

Substitutions are applied to clauses to produce specializations. The specialization of a clause is the result of the simultaneous replacement of every occurrence of each variable by the corresponding terms found in the pairs of the substitution. For example, given the program clause:

$$Q = p(X, Y) \leftarrow \top \wedge q(X) \wedge q(Y)$$

This clause can be specialized using the substitution θ given above. The specialized clause, denoted by $Q\theta$, is the clause:

$$Q\theta = p(a, X) \leftarrow \top \wedge q(a) \wedge q(X)$$

3.1 AGENTS FOR KNOWLEDGE MODELLING

Substitutions can also be applied to terms and to positive basic literals.

The *composition* of two substitutions is defined as follows. Let $\alpha = \{(X_1^1, t_1^1), \ldots, (X_n^1, t_n^1)\}$ and $\beta = \{(X_1^2, t_1^2), \ldots, (X_m^2, t_m^2)\}$. The composition $\alpha\beta$ is obtained deleting from the set $\{(X_1^1, t_1^1\beta), \ldots, (X_n^1, t_n^1\beta), (X_1^2, t_1^2), \ldots, (X_m^2, t_m^2)\}$ the pairs $(X_i^1, t_i^1\beta)$ such that $X_i^1 = t_i^1\beta$ and the pairs (X_j^2, t_j^2) such that $X_j^2 = X_i^1$ for some value of $i \in \{1, \ldots, n\}$.

A substitution θ is *more general* than a substitution α if there exists a third substitution β such that $\alpha = \theta\beta$.

To illustrate these definitions, let us consider the following substitutions:

- $\theta = \{(X, f(Y)), (Y, g(Y, Z))\}$;

- $\alpha = \{(X, f(a)), (Y, g(a, f(b)))\}$;

- $\beta = \{(Y, a), (Z, f(b))\}$.

It is not difficult to check, using the definitions above, that $\alpha = \theta\beta$. Therefore, θ is more general than α. Let us consider again the clause $Q = p(X, Y) \leftarrow \top \land q(X) \land q(Y)$. We can build the specializations $Q\theta$ and $Q\alpha$:

- $Q\theta = p(f(Y), g(Y, Z)) \leftarrow \top \land q(f(Y)) \land q(g(Y, Z))$;

- $Q\alpha = p(f(a), g(a, f(b))) \leftarrow \top \land q(f(a)) \land q(g(a, f(b)))$.

Clause Q characterizes the connection between relations p and q. Clause $Q\theta$ is a specialization of Q, which characterizes a connection between p and q that works only for those values of X and Y that are related via the functions f and g. Clause $Q\alpha$ is also a specialization of Q, which characterizes a connection between specific values that can be assigned to X and Y.

Clause Q is more general than $Q\theta$. Since $Q\alpha = (Q\theta)\beta$, the clause $Q\theta$ is more general than $Q\alpha$.

Let us consider now the following positive basic literals:

- $P_1 = p(X, f(Y, g(a, b, Z)))$;

- $P_2 = p(g(U, W, c), f(V, g(U, W, c)))$

in which X, Y, Z, U, V, W are variables, a, b, c are constants and f, g are functions.

The substitutions:

- $\theta = \{(X, g(a, b, c)), (Y, V), (Z, c), (U, a), (W, b)\}$;

- $\alpha = \{(X, g(a, b, c)), (Y, c), (Z, c), (U, a), (V, c), (W, b)\}$; and

- $\beta = \{(X, g(a, b, c)), (V, Y), (Z, c), (U, a), (W, b)\}$

have a common feature: if any of them is applied to the two positive basic literals above, the resulting specializations are identical:

- $P_1\theta = P_2\theta = p(g(a, b, c), f(V, g(a, b, c)))$;

- $P_1\alpha = P_2\alpha = p(g(a, b, c), f(c, g(a, b, c)))$;

- $P_1\beta = P_2\beta = p(g(a, b, c), f(Y, g(a, b, c)))$.

If a substitution produces identical specializations for two positive basic literals, then it is named a *unification* of those literals.

Considering the unifications above, we have that θ is more general than α and β and that β is more general than α and θ.

If two substitutions σ_1 and σ_2 are such that σ_1 is more general than σ_2 and σ_2 is more general than σ_1 they are called *equivalent*

3.1 AGENTS FOR KNOWLEDGE MODELLING

substitutions. It can be proved that for any substitutions β_1 and β_2, if $\sigma_1 = \beta_1 \sigma_2$ and $\sigma_2 = \beta_2 \sigma_1$ then these substitutions are formed by pairs (X_i, Y_i) in which X_i and Y_i are variables.

In Robinson (1965) we find an interesting and important result for logic programming:

> Given two arbitrary positive basic literals, if they are unifiable then there is a single set of equivalent unifications that is more general than any other unification. There is also an algorithm to build an arbitrary element of this set, or to answer no *if the literals are not unifiable*.

In Apt (1994) we find a non-deterministic version of this algorithm, as presented in Figure 3.1.

Let us consider now a set of program clauses as below:

$$P = \left\{ \begin{array}{c} C_1 = p(X,Y) \leftarrow \top \wedge q(X) \wedge q(Y) \\ C_2 = q(X) \leftarrow \top \wedge r(X) \\ C_3 = r(a) \leftarrow \top \\ C_4 = r(b) \leftarrow \top \end{array} \right\}$$

What values of X belong to the relation p, given the value b for Y? The answer to this question can be obtained as shown in Figure 3.2.

The value b for X is, therefore, one of the values that solves the problem. The value a also solves this problem, and it can be obtained using clause C_3 and the unification $\alpha = \{(X, a)\}$.

This problem-solving method, formed by the interleaving of unifications and replacements, is called *resolution*. Formally, the resolution method can be presented as shown in Figure 3.3.

Detecting the impossibility to produce the clause $\bot \leftarrow \top$ is equivalent to the halting problem, hence it is undecidable. There are, however, classes of logic programs for which this detection is decidable.

Let $p(t_1, \ldots, t_n)$ and $p(u_1, \ldots, u_n)$ be positive basic literals. This algorithm takes as input the set of equations $t_1 = u_1, \ldots, t_n = u_n$. The goal is to answer *no* if they are not unifiable, or to build the unification θ belonging to the set of most general unifications as presented above. Initially, $\theta = \{\ \}$:

1. non-deterministically select an equation;

2. if the equation has the form $f(r_1, \ldots, r_m) = f(s_1, \ldots, s_m)$, then delete this equation and add the equations $r_1 = s_1, \ldots, r_m = s_m$;

3. if the equation has the form $c = c$, then just delete the equation;

4. if the equation has the form $f(r_1, \ldots, r_m) = c$, $c = f(r_1, \ldots, r_m)$ or $f(r_1, \ldots, r_m) = g(s_1, \ldots, s_k)$, then answer *no* and abort execution;

5. if the equation has the form $X = X$, then just delete the equation;

6. if the equation has the form $t = X$, in which t is not a variable, then replace the equation by $X = t$;

7. if the equation has the form $X = t$ and X occurs in t (e.g., if $t = f(X)$), then answer *no* and abort execution;

8. if the equation has the form $X = t$ and X does not occur in t, then add the pair (X, t) in θ and replace every occurrence of X by t in the remaining equations.

Figure 3.1 Unification algorithm.

3.1 AGENTS FOR KNOWLEDGE MODELLING

1. Given the positive basic literal $p(X,b)$, there exists a unification θ for $p(X,b)$ and $p(X,Y)$. Using the algorithm above, we obtain $\theta = \{(Y,b)\}$.

2. Applying θ to C_1, we obtain $C_1\theta = p(X,b) \leftarrow \top \wedge q(X) \wedge q(b)$. The pair (X,b) belongs to the relation p if the conjunction $\top \wedge q(X) \wedge q(b)$ is satisfied. This conjunction is satisfied if b belongs to the relation q and there is at least one value for X that also belongs to q. This value can be b or some other value.

3. Taking the clause C_2, we have that any value of X that belongs to r also belongs to q. Let us consider the substitution $\theta' = \{(X,b)\}$, equivalent to θ. This substitution is a unification of $r(X)$ and $r(b)$. Using this unification, we obtain the specialization $C_2\theta' = q(b) \leftarrow \top \wedge r(b)$.

4. From $C_1 = p(X,Y) \leftarrow \top \wedge q(X) \wedge q(Y)$, we get $C_1\theta = p(X,b) \leftarrow \top \wedge q(X) \wedge q(b)$. If we replace (based on C_2) $q(X)$ by $\top \wedge r(X)$, we obtain another form of specialization, by restricting the values of q to those that are also values of r. This leads to the construction of the clause $C_{12}\theta = p(X,b) \leftarrow \top \wedge r(X) \wedge q(b)$, in which the repetition of the constant $\top$ has already been "simplified".

5. Applying now θ' to $C_{12}\theta$, we obtain $C_{12}\theta\theta' = p(b,b) \leftarrow \top \wedge r(b) \wedge q(b)$. Applying θ' to C_2, we obtain $C_2\theta' = q(b) \leftarrow \top \wedge r(b)$. We can replace $q(b)$ by $r(b)$ in $C_{12}\theta\theta'$, thus obtaining $C_{122}\theta\theta' = p(b,b) \leftarrow \top \wedge r(b)$.

6. Taking the clause C_4 and performing a replacement as above, we finally get the clause $C_{1224}\theta\theta' = p(b,b) \leftarrow \top$.

Figure 3.2 Answering a query using program clauses.

> Given a set of program clauses $P = \{C_1, ..., C_n\}$ and a Horn query Q:
>
> - use a selection rule E_1 to select a positive basic literal $\varphi_q \in Q$;
>
> - use a selection rule E_2 to select a program clause $C_i \in P$, such that $C_i = \varphi_i \leftarrow \top \wedge \Psi_i$ and there exists a unification θ for φ_q and φ_i;
>
> - replace Q by $Q\theta$ and $\varphi_q\theta$ by $\Psi_i\theta$ in $Q\theta$;
>
> - repeat these steps, until the clause $\bot \leftarrow \top$ is produced or the impossibility to produce this clause is detected.

Figure 3.3 Resolution.

The clause $\bot \leftarrow \top$ denotes the existence of at least one set of values for the variables in Q such that the conjunction represented in Q does not conflict with the program clauses in P. This set of values is determined by the composition of unifications used in the production of $\bot \leftarrow \top$.

The *meaning* of a logic program is defined as the smallest sets of tuples of values that necessarily belong to the relations represented in the program, in order to ensure the satisfaction of every clause in the program. A program clause $\varphi \leftarrow \top \wedge \Psi$ is satisfied by tuples of values if, when the variables in the clause are replaced by the corresponding values in the tuples, whenever the conjunction $\top \wedge \Psi$ is such that each tuple of values belongs to the relation represented by Ψ, the tuple of values in φ also belongs to φ.

The last paragraph was mathematically precise, but rather cryptic. Let us clarify it through two simple examples. Let us

3.1 AGENTS FOR KNOWLEDGE MODELLING

first consider the logic program formed by the following four clauses:

1. $p(X) \leftarrow \top \land q(X) \land r(X);$
2. $q(a) \leftarrow \top;$
3. $r(a) \leftarrow \top;$
4. $r(b) \leftarrow \top.$

If a does not belong to q, then clause (2) is not satisfied. If a and b do not belong to r, then clauses (3) and (4) are not satisfied.

The value a is the only value that belongs to q AND r, hence it is the only value that must belong to p, so that clause (1) is satisfied.

This program therefore means that the following sets of values are associated with each relation in the program:

- $p: \{a\};$
- $q: \{a\};$
- $r: \{a, b\}.$

Let us now consider the logic program formed by the following three clauses:

1. $p(X, Y) \leftarrow \top \land q(X) \land q(Y);$
2. $q(a) \leftarrow \top;$
3. $q(b) \leftarrow \top.$

If a and b do not belong to q, then clauses (2) and (3) are not satisfied.

Taking clause (1) into account, if X and Y are replaced by a, then the pair (a, a) must necessarily belong to p so that this clause is satisfied. The same happens for every other combination of the values a and b.

This program therefore means that the following sets of values are associated with each relation in the program:

- $p: \{(a,a), (a,b), (b,a), (b,b)\}$;

- $q: \{a, b\}$.

There is a procedure to build the meaning of any logic program. This procedure is actually very simple, as we show below:

1. First, assign the empty set to each relation in the program.

2. Check the clauses one by one with respect to satisfaction. If a clause is not satisfied, then add the smallest set of values to the relation *to the left* of the implication $\leftarrow$ to satisfy the clause. Notice that, whenever a value is added to a relation in order to satisfy a clause, other clauses that were previously satisfied can become unsatisfied.

3. Iterate this process until all clauses are satisfied.

Consider the following logic program:

1. $p(X, Y) \leftarrow \top \wedge q(X) \wedge q(Y)$;

2. $q(X) \leftarrow \top \wedge r(X)$;

3. $r(a) \leftarrow \top$;

4. $r(b) \leftarrow \top$.

3.1 AGENTS FOR KNOWLEDGE MODELLING

Considering that initially the relations p, q and r have as their meanings the empty set, we have:

- $p: \{ \ \}$;

- $q: \{ \ \}$;

- $r: \{ \ \}$.

This meaning satisfies clauses (1) and (2), but clauses (3) and (4) are not satisfied. To satisfy clauses (3) and (4), we must modify the meaning of r as follows:

- $p: \{ \ \}$;

- $q: \{ \ \}$;

- $r: \{a, b\}$.

Now, clauses (1), (3) and (4) are satisfied, but clause (2), which was satisfied before, is no longer satisfied. To satisfy clause (2), we must modify the meaning of q:

- $p: \{ \ \}$;

- $q: \{a, b\}$;

- $r: \{a, b\}$.

Now clauses (2), (3) and (4) are satisfied, but clause (1) is not satisfied. To satisfy clause (1), we must finally modify the meaning of p:

- $p: \{(a,a), (a,b), (b,a), (b,b)\}$;

- $q: \{a, b\}$;

- $r: \{a, b\}$.

This procedure to build the meaning of the relations in a logic program is monotonic: a value is never deleted from the meaning of a relation. As presented in Apt (1994), based on a result by Alfred Tarski (1955), the meanings of relations built this way are minimal and correspond precisely to the values assigned to variables when a clause of the form $\bot \leftarrow \top$ is built based on a logic program and a Horn query.

Let $\mathcal{P}$ represent the set of all attributes. A *state* is the instantiation of all elements of $\mathcal{P}$.

An *action* is the change of value of an attribute. In other words, an action moves one attribute from one instantiation to another. As a consequence, an action characterizes the transition between two states.[1]

A *goal* is just some particular state. Goals are generated by *motivations*. Let us denote as $\mathcal{S}$ the set of states, $\mathcal{M}$ the set of motivations and $\mathcal{G} \subseteq \mathcal{S}$ the set of goals. For each particular state, different motivations determine different sets of goals. Formally, this is represented by *goal generating functions* $g: \mathcal{S} \to (\mathcal{M} \to 2^{\mathcal{G}})$. Given a specific state $s \in \mathcal{S}$, we have a functional relation between motivations and sets of goals.

Actions, motivations and goal generating functions are also assumed to be independent of each other. If necessary, we can also build *clausal theories of actions, motivations and goal generating functions*.

An action can be represented as a ternary predicate $a_i(att_i, vi_i, vf_i)$, in which att_i is the name of an attribute, vi_i is

---

[1] Strictly speaking, an action is *represented* by the change of value of an attribute.

3.1 AGENTS FOR KNOWLEDGE MODELLING

the initial value of the attribute (i.e., the value of att_i before the action is executed) and vf_i is the final value of the attribute (i.e., the value of att_i after the execution of a_i).

A goal generating function is slightly more complicated to represent in a form that suits the construction of clausal theories. Since it is a function, its representation must be done as a collection of predicates of the form $g(s, mg)$, which are the tabular representation of the goal generating function g. Each state s is a list of pairs (att_i, val_i), in which att_i is the name of an attribute and val_i is a corresponding value. Given a certain state s, the function g generates a *function* from motivations to sets of goals. Hence, the term mg must be the tabular representation of this function. It can be a list of pairs (mot_i, gs_i), in which mot_i is the name of a motivation and gs_i is a list of states, which are precisely the goal states generated by mot_i.

An *agency* is characterized by a non-empty set of instantiated attributes, a set of actions, a set of motivations and a set of goal generating functions.

Elements of each of these sets can be *persistent* or *non-persistent*:

- A persistent instantiated attribute cannot change, otherwise the agency is transformed into another agency. Non-persistent attributes can change their instances. If an agency has a non-persistent attribute characterizing it, then it also has a subset of the type of that attribute that characterizes the *allowed instances* of the attribute for that agency. As an example, let us consider we are trying to characterize a banana as an agency. The attribute p_1 representing "is a fruit", has type $\{yes, no\}$. For a banana, it *has* to be instantiated as $p_1(yes)$. The attribute p_2 representing "colour", however, has as its type the myriad of all conceivable colours. The allowed instances of p_2 for a banana are, however, $\{green, yellow, black\}$. Clearly, a banana changes from green to yellow as it ripens, and then from yellow to black as it rots. The banana does not cease to be a banana by changing from green to yellow to black. It would however no longer be the original agency if it changed to blue, or if it ceased to be a fruit. The attribute p_1 is a persistent one, whereas p_2 is a non-persistent attribute.

- A persistent action is one that the agency *can* always execute. A non-persistent action is a skill that can be acquired or forgotten by the agency. For example, a living dog can breathe, but the day it loses this capability it ceases to be a living dog. A living dog, immediately after birth, however, cannot run. It acquires this capability with time, and later in life it may also lose this capability, yet it is still a living dog all the way through. We denote the set of actions associated with an agency as C – the *capabilities* of the agency.

- Similarly, a persistent motivation is one that is always present within the agency. Our living dog is always driven by a *life preservation instinct*, for example. A non-persistent motivation can come and go (e.g., *hunger*).

- Finally, given a state and a motivation, a persistent goal generating function always generates the same set of goals, whereas a non-persistent goal generating function can generate different sets of goals at different moments.

Given an agency Ag_1, we can build an agency Ag_2 by adding some elements to any of the sets above that characterize Ag_1. We call this the *specialization* of an agency. Specialization is a powerful conceptual tool, which will be thoroughly used in the remainder of this text.

An *autonomous agency* adds to the concept of agency above the idea of *perception* and the related concept of *beliefs*. *Perception* is a mapping from states to states, leading from the actual state to some state perceived by the agency. The perceived state does not necessarily agree with the actual state: they may contain different attributes, different instances for common attributes, and the

3.1 AGENTS FOR KNOWLEDGE MODELLING

quantity of attributes may differ between the perceived and the actual state.

The perceived state is the state the agency *believes to be the actual one*. Each agency has a unique and personalized perception mapping $b: S \to S$. Perceptions are non-persistent, and typically the perceptions of an agency are refined with time, approximating the perceived states to the actual ones – although this gradual approximation must not be regarded as a necessary condition for every agency.

An autonomous agency acts according to its own perceptions (and hence beliefs). Let b^S denote the set of perceptions of an agency. Then, the *autonomous goal generating function ag* determines a set of goals given the motivations of the agency and its *perceived state*, based on its *beliefs*: $ag : b^S \to (\mathcal{M} \to 2^{\mathcal{G}})$.

So far, however, we have not stated any explicit relationship between actions and goals. The final ingredient to characterize an agency is, therefore, the capability to devise *plans* that are sequences of interleaved actions and perceptions aiming at changing present (perceived) states to goals. More precisely, a plan is a directed graph, in which nodes represent actions and perceptions, and the directed edges determine the sequence in which actions and perceptions should occur. A plan is generated given a set of goals and a perceived state. Thus, an autonomous agency also contains an *autonomous plan generating function* $ap: b^S \to (2^{\mathcal{G}} \to graph(\mathcal{C}))$, where $graph(\mathcal{C})$ is a directed graph as above. We assume that *perceive* is an *action* $\in \mathcal{C}$.

This characterization suffices to describe an isolated agency. The knowledge of an agency, as will unfold in the remainder of this text, is expressed in the plans it devises; that is, given the motivations and the actual perceived state of an agency – and thus given the actual set of goals acknowledged by that agency – the agency devises a sequence of actions that will take it from its actual state to some of its goals. The capability of the agency that makes it possible for these plans to be devised is a direct consequence of the *knowledge* it has.

If $\varphi \in b^S$ for an agency i, then we say that agency i *knows about* φ and denote this as $K_i\varphi$. Together with the motivations of agency i, φ influences determination of the goals of this agency.

Cooperative (i.e., non-isolated) agencies are agencies that acknowledge and count on resources from other agencies. For an agency to cooperate with another agency, it must be persuaded to do so. This is clearly a complex and subtle problem. We are going to touch on this problem at some points in this text, but the focus here is on the perception of mutual useful resources. Evidently, the problem of "persuading" a software agency to cooperate with another agency is infinitely simpler than the problem of persuading human agents to cooperate with agencies.

Two possibilities arise at this point: one agency may be interested in other agencies' *knowledge bases* or in other agencies' *capabilities*. The former gives rise to the discipline called *knowledge sharing*, a central issue for codification-based knowledge management. Here, the core technology in use has been *artificial ontologies*, and this is the subject of Chapter 4. The latter (other agencies' *capabilities*) is the foundation of knowledge coordination based on *capabilities sharing*, which is a central issue for personalization-based knowledge management. This topic is discussed in Chapter 5.

Before we close this section, we need to relate this characterization of agencies with the discussion in Chapter 2 about data and information.

Our agencies operate at the semantic level (i.e., they live in the "real" world rather than the "symbolic" world). In other words, attributes are extracted from the set of *referents* that can be used to interpret symbolic data, as proposed in Chapter 2. Perceptions are based on these referents, as are motivations, goals, actions and plans.

Our perfect agency is one with total, complete, bijective data interpretation with the *identity mapping* as the perception mapping. An *identity* perception mapping ensures perfect appraisal of reality, and the other properties ensure that the activities and the knowledge of the agency can be captured symbolically to their full extent. This is scarcely achieved even for designed artificial agencies.

Incomplete interpretations give room to what has been called by the knowledge management community *tacit knowledge*. Tacit knowledge is, therefore, that portion of information that does not admit symbolic representation.

Partial interpretations are, in some sense, the "dual" of incomplete interpretations. Partial interpretations characterize partial ignorance of the agency about occurrences in the world. Since these occurrences are recorded in the form of data, they can be accessed by other agencies. Partial interpretations therefore characterize deficiencies in the knowledge of an agency that should be remedied through *learning*.

Inconsistent and redundant interpretations characterize defective data-processing systems. Redundancies trigger more than one data update for a single change in the real world, and inconsistencies disrupt the reliability of data.

Organizational modelling and design should strive for the best approximations of total, complete, bijective interpretations and perception systems as close as possible to identity mappings. We should be aware, however, that imperfections exist and that our models should be prepared to cope with these imperfections.

3.2 AGENTS FOR ORGANIZATIONAL MODELLING AND DESIGN

According to Bernus and Nemes (1999):

> *organisational design is the art of creating agents out of agents.*

We propose the characterization of agencies given above as the basis for organizational modelling and design. To support this proposal, we reconstruct the relevant concepts of the main management schools in terms of agencies.

3.2.1 Agencies and knowledge in the different schools of management

The classical school of management was based on expectations of stability: transient states of instability should be followed by lengthy steady states. Hence, agencies could and should be designed to function at those steady states, in which the values of most attributes should be utterly predictable.

Agencies were modelled according to simplified and mechanistic views. Most often, the single *motivation* taken into account was called *perfect rationality*, which meant the desire to maximize financial reward.[2]

Perceptions and actions were stereotyped for every agency, and individual variations were not considered or, if that was not possible, deemed harmful and disruptive. The same occurred for *goal generating* and for *plan generating* functions.

From the classical point of view, the ultimate goal of an agency should be to function as a deterministic machine. Managers should not try to analyse, exploit or give value to motivations and knowledge of their workers, just as a carpenter does not consider the motivations and knowledge of a hammer.

Chester Irving Barnard advanced many concepts employed nowadays in knowledge management, and in the following paragraphs we detail how his organizational model comprises these concepts.

We start by rephrasing Barnard's original definition of formal

---

[2] We disagree profoundly with calling this *perfect rationality*. We leave a thorough discussion of this point to our forthcoming companion publication on knowledge, existence, truth and reasoning.

3.2 AGENTS FOR ORGANIZATIONAL MODELLING AND DESIGN

organization, to become *conscious, deliberate and purposeful co-operation structures among agents*.

Among the different actions associated with agencies, we have *information-based actions*, which are messages, requests and replies, orders, reports, etc. Information-based actions do not directly change the states of the physical world. They may nevertheless induce such changes (e.g., by ordering an agency to perform a world-changing action, or by generating a change in the beliefs of an agency that eventually leads to an action that changes the physical world).

Actions – information based or otherwise – occur via pre-determined *action channels*, that relate each action to its corresponding subjects. Action channels have as a special case *communication channels*: a communication channel is the action channel of an information-based action.

An action is *effective* in the sense of Barnard if for an agency it contributes to achieving a goal. It is *efficient* if it does not deter any agency (i.e., the performer of the action and all other agencies) from achieving any goal.

An ideally well-designed agency is such that the set of actions linked to all participating agencies is optimized in terms of effectiveness and efficiency. Note that this is a necessary and sufficient condition, if *all* agencies are taken into account and if every agency *knows* that its corresponding actions are indeed effective and efficient with respect to its own goals and that other agencies' actions are also efficient with its goals. This accounts for Barnard's objective and subjective incentives.

An organization can be characterized as a hierarchy of agencies, together with a carefully designed structure of action channels. *Management thus becomes the activity of permanently supervising and updating the organization, adapting its design in such way as to bring it as close as possible to the ideal of an optimally designed organization.*

The foundation of the socio-technical school is general systems theory. In the socio-technical school we find emphasis on the diversity of agencies, in clear opposition to the classical school. Here, motivations are considered to be fluid, dynamic, manifold and specific to each agency.

This view is compatible with our characterization of agencies based on attributes. By removing one or more attributes to the definition of an agency, we characterize a different – more general – agency. For example, we can have a whole department in an organization behaving as an agency and each individual working for that department being a specialization of the department. Agencies at all levels in the hierarchy (e.g., departments and individuals) each have motivations of their own.

Different agencies also have different capabilities, goal generating functions, plan generating functions and perceptions. All these features are also fluid and dynamic, as proposed in the previous section. As a logical consequence, *states* must also have ever-changing features.

In Hansen et al. (1999) we find two alternative strategies for knowledge management, coined *codification strategy* and *personalization strategy*.

The codification strategy relies on explicit (i.e., symbolic) representations of knowledge. Knowledge is extracted from agencies and stored using some sort of linguistic representation (paper documents, computer-based information, etc.). Once it is done, knowledge becomes independent of the originating agencies, and it can be reused by other agencies without direct intervention from the original sources of that knowledge.

The personalization strategy, on the other hand, relies on the construction of networks of capabilities – frequently called *knowledge maps* – that identify for each piece of knowledge and for each conceivable activity the agencies prepared to provide the organization with it.

The codification strategy brings to the knowledge-based organization the same potential benefits that Taylor and Ford brought to manufacturing. It is also applicable to situations similar to those in which those models were successful. As

3.2 AGENTS FOR ORGANIZATIONAL MODELLING AND DESIGN

pointed out in Hansen et al. (1999), the codification strategy is advantageous for agencies dealing with standardized products and stable markets. It is also applicable for mature products (i.e., products whose technology is well known) such that the corresponding knowledge about its manufacturing and marketing can be conveniently represented symbolically. As pointed out in Starkey (1992), this is the scenario for which the classical management models were developed.

The personalization strategy is advised for the opposite situation; that is, agencies dealing with customized and/or innovative products, dynamic markets and situations in which the relevant knowledge for the organization may not be fully represented (as mentioned before, many authors call *tacit knowledge* those pieces of information that are relevant for an organization, but which we do not know how to extract from an agency or how to represent them symbolically).

Looking back at Barnard's model, we see that this dichotomy had also been identified by that author.

Important corporations have employed these two strategies with success, although it has been observed that these strategies tend to be mutually exclusive. As we find in Hansen et al. (1999), companies have adopted the codification strategy:

- Andersen Consulting, Ernst & Young and Dell.

On the other hand, the following companies have adopted the personalization strategy:

- Bain, Boston Consulting Group, McKinsey and Hewlett-Packard.

The system of incentives must target different behaviours depending on the strategy adopted. The codification strategy must bring to agencies incentives to actions that make their *knowledge* explicit and available to other agencies, independently of their direct intervention through actions. The personalization strategy, on the other hand, must bring incentives for agencies to make their *actions* available to other agencies, without necessarily making their knowledge available through linguistic codification.

In the following chapters we discuss conceptual tools for each strategy in turn. In Chapter 4 we analyse *artificial ontologies*, an important conceptual tool originated from information technology that has had widespread use and has been extensively advertised as the solution for (codification-based) knowledge management. The basic proposal of artificial ontologies is that the capabilities of an agency can be expanded via shared beliefs/ perceptions of other agencies, encoded as corporative "knowledge bases". It is the *know about* approach for knowledge coordination, in which agencies ask each other "may I borrow your knowledge?"

In Chapter 5 we propose an original conceptual tool, which we call *structures of capability providers*, and show how it can be used to support the design and maintenance of agencies regarding knowledge coordination based on a personalization strategy. The personalization strategy suggests that agencies should expand their own capabilities with other agencies' capabilities and actions. This is the *know how* approach, in which agencies ask each other "may I borrow your skills?"

REFERENCES

Apt, K. R. (1994) "Logic programming," in J. van Leeuwen (ed.), *Handbook of Theoretical Computer Science*, Elsevier–MIT Press.

Bernus, P. and Nemes, L. (1999) "Organisational design: Dynamically creating and sustaining integrated virtual enterprises," *Proceedings of IFAC World Congress, Beijing*.

REFERENCES

Carley, K. M. (1995) "Computational and mathematical organization theory: Perspective and directions," *Computational and Mathematical Organization Theory*, **1**(1).

Fisher, M., Muller, J., Schroeder, M., Stanford, G. and Wagner, G. (1997) "Methodological foundations for agent-based systems," *The Knowledge Engineering Review*, **12**(3), 323–329.

Hansen, M. T., Nohria, N. and Tierney, T. (1999) "What's your strategy for managing knowledge?," *Harvard Business Review*, **106**, March–April.

d'Inverno, M. and Luck, M. (2001) *Understanding Agent Systems*, Springer-Verlag.

d'Inverno, M., Fisher, M., Lomuscio, A., Luck, M., de Rijke, M., Ryan, M. and Wooldridge, M. (1997) "Formalisms for multi-agent systems," *The Knowledge Engineering Review*, **12**(3), 315–321.

Luck, M. and d'Inverno, M. (2001) "A conceptual framework for agent definition and development," *The Computer Journal*, **44**(1), 1–20.

Mendelson, E. (1987) *Introduction to Mathematical Logic*, Wadsworth & Brooks/Cole (3rd edition).

Muller, J. P. (1998) "Architectures and applications of intelligent agents: A survey," *The Knowledge Engineering Review*, **13**(4), 353–380.

Prietula, M. J., Carley, K. M. and Gasser, L. (eds) (1998) *Simulating Organizations: Computational Models of Institutions and Groups*, AAAI/MIT Press.

Robinson, J. A. (1965) "A machine-oriented logic based on the resolution principle," *Journal of the ACM*, **12**(1).

Russell, S. and Norvig, P. (1995) *Artificial Intelligence – A Modern Aproach*, Prentice Hall.

Shoenfield, J. (1967) *Mathematical Logic*, Addison-Wesley.

Starkey, K. (ed.) (1992) *How Organizations Learn*, Thomson Business Press.

Tarski, A. (1955) A lattice-theoretic fix-point theorem and its applications. *Pacific Journal of Mathematics*, **5**, 285–309.

4
Ontologies

> *On themes that are common to us all I shall not speak in the common language; I am not going to repeat you, comrades, I am going to dispute with you.* – Boris Pasternak

In Chapter 2 we analysed the concept of *knowledge*, and in Chapter 3 we proposed *agencies* as the basic conceptual tool to model organizations and their management.

Knowledge management is reputed to be a key issue in modern organizational management (Davenport and Prusak, 1998; Nonaka, 1991). As pointed out in Chapter 2, the essence of knowledge management is how agencies do the following:

- Generate and acquire knowledge – this relates to how agencies measure and foster productivity in innovation and creativity, identify weaknesses in their competences and decide how to correct them (e.g., by hiring experts in appropriate fields, training their personnel, incorporating novel technologies, revising incentive systems, etc.).

Knowledge Coordination   F. S. Corrêa da Silva and J. Agustí-Cullell
© 2003 John Wiley & Sons, Ltd   ISBN: 0-470-85832-X

- Store and preserve knowledge (often quoted as "organizational learning", see Garvin, 1993; Senge, 1990; Starkey, 1992) – this relates to how knowledge can be preserved within agencies once it has been obtained.

- Access and use knowledge – this relates to how agencies identify relevant pieces of knowledge when facing new situations and challenges and what is needed to build a structure such that those pieces of knowledge can be efficiently retrieved when necessary.

- Distribute and disseminate knowledge – this is based on the assumption that knowledge is distributed across agencies. Hence, different parts of an agency hold different skills and capabilities. This adds a new dimension to the access and use of knowledge, requiring components of an agency to be capable of communicating with each other, expressing capabilities they may need, problems they may be interested in delegating to other agencies, as well as solutions to delegated problems.

Knowledge distribution and dissemination is important, complex and multifaceted. Among complicating factors for knowledge distribution and dissemination is the heterogeneity of agencies: agencies can be natural or artificial, agencies can be composed of other agencies, agencies have different capabilities and capabilities can be extremely dynamic.

For agencies to communicate they must understand each other. More specifically, agencies must be able to understand and make good use of each other's *knowledge*. Furthermore, they must be able to identify correspondences between their own knowledge and the knowledge of other agencies, as well as between tasks to be executed and capabilities that allow their execution. Researchers in information technology have considered the problem of the communication among agencies quite thoroughly, especially for communication among artificial software agencies (see, e.g., Corrêa da Silva and Meneses, 2001, 2002; Kinny, 2001; Klusch, 2001).

Effective communication requires that different agencies commit to a shared terminology. This has been misnamed *shared ontology*. In the following sections we present why we believe this is a misnomer and identify the proper role of shared terminologies for communication among agencies (including natural agencies). We concede to *artificial ontology* instead of "terminology" throughout this work due to the popularity gained by this term.

4.1 ONTOLOGIES – NATURAL AND ARTIFICIAL

In the *Merriam-Webster Collegiate English Dictionary* we find the following entry for "ontology":

> Function: noun
> Etymology: New Latin *ontologia*, from ont- + -logia -logy
> Date: circa 1721
> 1 : *a branch of metaphysics concerned with the nature and relations of being*
> 2 : *a particular theory about the nature of being or the kinds of existents*

Ontology is therefore a branch of metaphysics, and the term itself has existed since at least the early eighteenth century.[1]

The subject matter of ontology is existence. Existence can be considered from absolute, subjective or social viewpoints:

---

[1] This is one reason that makes us feel uncomfortable when we find highly reputed researchers in information technology claiming that "Researchers in artificial intelligence first developed ontologies to facilitate knowledge sharing and reuse" (Ding et al., 2002; Fensel, 2001; Fensel et al., 2001).

- *absolute existence* discusses the possibility of being of an entity irrespective of that entity being perceived by another entity;

- *subjective existence* discusses the nature and relations of being relative to an agency; and

- *social existence* discusses the relations of being as perceived collectively by many agencies.

Committing to an ontology makes sense from the last two viewpoints. From the subjective point of view, if an agency commits to an ontology, it means that this agency is accepting the existence of certain entities and the proposed relations among them. From the social point of view, two or more agencies can (partially) commit to a single ontology, thus agreeing upon the existence of entities and relations among them.

Commitment to a single ontology is a prerequisite for communication to occur. Requiring (or expecting) a human agent to commit to an ontology almost certainly implies the expectation that this agent accedes to an understanding of abstract as well as concrete entities, whose symbolic representation may or may not be possible. Ontological commitment is therefore a hard, time-consuming activity that defies formalization.

Commitment to a single ontology by many agencies inherits and intensifies the difficulties of ontological commitment for a single agency. Typically, it is the result of social interaction and willingness to cooperate and to communicate: agencies engage in deliberate mutual effort to bridge their assumptions about the environment and about themselves, to avoid being solitary.[2]

---

[2] It is worth remarking that ontological commitment does not require *agreement*: two persons may commit to a single ontology for "God", for example, even if one is an atheist and the other is a religious person. As a curious corollary to this remark, we cannot be an atheist if we do not acknowledge the existence of "God" as a concept.

We claim that a knowledge management model for agencies involving human agents that requires the commitment of all components to a single ontology is founded on resources that will hardly be available in realistic situations.[3]

For agencies in which all agents are artificial and have been designed under normative principles, things become easier. The ontologies to which these agencies commit are also artificial and, since these agencies can be described symbolically in every detail, so might their ontologies be likewise expressed.

It is interesting to note that the classical school of management is also normative, even though it deals with people instead of artificial software agencies. Among the classical proposals, we find those based on bureaucratic concepts, which in turn were founded on *division and specialization of work*, *hierarchy of authority*, *rationality* (i.e., optimized utilization of resources) and especially *explicit rules and patterns of behaviour*, *documentation* and *impersonality*. These foundations perfectly coincide with the hypotheses and requirements to use centralized artificial ontologies. For this reason, we classify management proposals based on commitment to centralized artificial ontologies as a *neo-bureaucratic movement*.

4.2 IMPLEMENTING AND USING ARTIFICIAL ONTOLOGIES

Artificial ontologies are structured tables encoding the underlying languages that represent artificial agencies and their

---

[3] We must however remark that this is not a consensual view. For example, in Fensel et al. (2001) we find: "Ontologies are becoming popular largely because of what they promise: a shared and common understanding that reaches across people and application systems." An almost literal copy of this sentence can also be found in Ding et al. (2002).

environments. A widely cited definition for artificial ontologies can be found in Gruber (1993):

An ontology is an explicit specification of a conceptualisation.

Only artificial agencies admit complete, explicit representations of their conceptualizations. Furthermore, only artificial agencies built upon normative theories of information, knowledge and action can have their conceptualizations *specified*.

Many systems and standards have been proposed for the development of artificial ontologies (see, e.g., `http://www.ontology.org` as an example of organization aiming at the standardization of artificial ontologies and their development methodology), thus giving rise to a new engineering discipline, frequently called *ontological engineering* (Fensel, 2001; Fensel et al., 2001). Recently, some successful applications have been reported based on information sharing among software systems based on partial commitment to standardized artificial ontologies (Kalfoglou et al., 2000).

Of special interest are the recent proposals to furnish World Wide Web (WWW) sites and pages with content based on shared artificial ontologies. As mentioned in Fensel (2001), the WWW contains around 300 million "static objects" providing all sorts of information. Furthermore, the widespread use of WWW architecture and protocols has made it a *de facto* standard for the design and construction of distributed, object-based information systems.

These "static objects" have been built for a variety of purposes, making use of diversified technological resources and immersed in different sociocultural and organizational contexts. All this diversity hinders the retrievability of useful information from the WWW and asks for some sort of indexing of information. Indeed, this is the service that many successful systems provide (e.g., those of Yahoo, Altavista, Google and Excite).

4.2 IMPLEMENTING AND USING ARTIFICIAL ONTOLOGIES

Methods and techniques to structure the content of WWW sites and pages have been proposed and developed around the Extensible Mark-up Language – XML (http://www.w3c.org/xml). XML is a mark-up language that supports the organization of taxonomies and terminologies. Hence, it is an appropriate tool for developers of Web-based information systems to encode and share terminological information.

If the content of the WWW were tagged with terminological information, smarter searches could be performed, thus making it more useful and effective. This possibility is of particular interest for bounded webs, or *intranets* (i.e., computer-based organizational information systems that employ WWW architecture and protocols for their implementation).

Another possibility is the development and use of tools developed specifically to deal with artificial ontologies. In Fensel (2001) we find pointers to a plethora of such systems, designed to support editing, merging and consulting artificial ontologies.

However, the practical applicability of standardized artificial ontologies in the large has been challenged by the research community in the field. As pointed out in Uschold (1998), many of the proposed systems have been experimentally applied to few agencies, a situation in which it may not be worth the effort to build artificial ontologies to implement information sharing.

Furthermore, in Corrêa da Silva et al. (1999, 2001) we have a systematic account of some technical impediments to a widespread use of artificial ontologies. These impediments are presented in rather technical terms through critical examples in which information sharing among artificial software agencies by means of artificial ontologies can be hard. Essentially, the problem is that the *origin* and the *resources needed* to synthesize a piece of information can determine whether it can be shared. Hence, information sharing is not only a matter of communication, but also of mutual acknowledgement of skills and capabilities – even for artificial agencies. Even in many situations in which artificial ontologies can be deemed necessary for cooperation among agencies, they may not be sufficient to ensure effective communication

In Chapter 2 we discussed the different forms of relationship between data and their referents, thus characterizing a dichotomy between data and information. We have also observed that some referents may not have corresponding data to represent them. As quoted above, ontologies are theories about kinds of existents. Hence, an ontology must also be at "the referent level", since this is the level where the existents are.

Artificial ontologies, on the other hand, are essentially symbolic. A piece of code containing an artificial ontology can be completely devoid of meaning to someone who has not been trained to use the specific editor of artificial ontologies that was used to write that piece of code, or to someone who is unaware of the context and desired interpretation of the symbols in that piece of code. Therefore, artificial ontologies are just collections of data, with no intrinsic meaning attached to them. This is the reason that we considered the term *ontology* a misnomer for this concept.

Given the appropriate context, artificial ontologies can be very useful. As we have tried to stress, however, the inappropriate use of artificial ontologies can vary from inneffective to utterly disastrous.

Although it is suggested that artificial ontologies can be used to encode the conceptualization of everything (including processes, capabilities and domain information), they have been used primarily to encode domain information. Thus, if it can be assumed by different agencies that their capabilities are equivalent and if it can also be assumed that every piece of relevant information for every agency is encoded in the "knowledge base" of some agency, then – and only then – artificial ontologies can be useful.

Artificial ontology editors and software for merging and aligning artificial ontologies, in this case, provide the appropriate tools for standardization of the metadata used to organize the

4.2 IMPLEMENTING AND USING ARTIFICIAL ONTOLOGIES

representation of the information contained in the different "knowledge bases". Indeed, the "knowledge base" of an agency is actually a distributed database, whose elements are the "knowledge bases" of the components of the agency. Merging of artificial ontologies is equivalent to fusion of the database schemata of different elements of a distributed database into a single schema. Aligning of artificial ontologies corresponds to the identification of conceptual mappings among those different database schemata.

Typically, an artificial ontology is implemented as a polymorphic classification of the relevant concepts occurring within a system: a hierarchy of concepts is organized as a directed acyclic graph with a single root element. Each node in the graph is determined by a set of *attributes*, and if there is a path between nodes n_i and n_j, then the collections of attributes A_i of n_i and A_j of n_j are such that $A_i \subseteq A_j$ (i.e., n_j is a *specialization* of n_i (resp., n_i is a *generalization* of n_j)).

Additional relationships betweens concepts can be defined and represented as a directed hypergraph in which nodes are the concepts above and arrows are relationships between the sets of concepts.

Given a symbolic system S and an artificial ontology $\mathcal{O}_S$ for S, each symbol comprising the language of S is classified as an *instance* of a node $n \in \mathcal{O}_S$, in which the attributes of n assume specific *values*. A set of *equations* determines interrelationships among attributes.

The structure of the acyclic graph induces a partial order on the nodes of an artificial ontology, with the root of the graph as the greatest element $\top$. The correspondence between two artificial ontologies $\mathcal{O}_1$ and $\mathcal{O}_2$ determines order relations between elements of those artificial ontologies, so that it makes sense to look for the *least upper bound* (resp., the *greatest lower bound*) of a node of $\mathcal{O}_i$ in $\mathcal{O}_j$, which is the *most specific concept* in $\mathcal{O}_j$ that is a generalization of the concept in $\mathcal{O}_i$ (resp., the *most general concept* in $\mathcal{O}_j$ that is a specialization of the concept in $\mathcal{O}_i$).

As presented in Hansen et al. (1999), the core concept in codification-based knowledge management is reuse, and the credo is that information reuse is best done by storing and retrieving information in some way that it becomes independent of its sources.

A sound metaphor for codification is that of the construction of a *knowledge library*. Codification-based knowledge coordination is akin to the work of a librarian, who is prepared to identify the needs of a user and to guide the user to the appropriate slot in the "knowledge library" (the analogue of a *book* in a conventional library) that contains the needed piece of information.

Artificial ontologies work as the indexing system for the library. They provide the navigational structure to find useful information within the library.

The best way to understand the underlying concepts of artificial ontologies is by analysing a well-designed example. In the following section we present two examples. The first one is the Resources–Events–Agents (REA) artificial ontology, which is a well-known and appraised artificial ontology for business processes. The second example is the Lattes integrated system for academic CVs used in Brazil. The REA artificial ontology has been developed for years and is a fine example of a carefully constructed artificial ontology, which nevertheless has been less used than we believe their proponents had expected. The Lattes system, on the other hand, had a very modest start and is now widely used by a large academic community in Brazil and some parts of Latin America.

4.3 ILLUSTRATIVE EXAMPLE I: THE RESOURCES–EVENTS–AGENTS ENTERPRISE ONTOLOGY

The REA artificial ontology is primarily the result of the work of a single researcher, W. E. McCarthy, who is a professor at

4.3 ILLUSTRATIVE EXAMPLE I

Michigan State University. As indicated by that author, its seminal paper is McCarthy (1982). Initially, it was proposed as the means to update accounting methods.

As pointed out in McCarthy and Geerts (1997), the practices of modern accounting are founded on the methodology proposed by Luca Pacioli, a Franciscan monk who lived in Italy during the 15th century. Accounting was designed to provide financial analysts with data about past events, something that is useful for auditing and for long-term strategic planning in stable economic scenarios.

Electronic business services using the Internet have brought the necessity of real-time financial control and support for remote transactions. This has changed the traditional supply chain model to significantly more complex distributed supply networks.

The REA model is a rational reconstruction of the procedures involved in supply networks, furnishing managers with real-time accounting information. Recently, it has been proposed as an artificial ontology of economic transactions, especially useful for the implementation of electronic business services (Geerts and McCarthy, 1999; Haugen and McCarthy, 2000).

An REA agent corresponds precisely to our notion of agency. A resource in the REA model is anything of value to an agency that is held by some of its components (e.g., cash, inventory and capabilities). An event is the transfer of control of a resource from one agency to another.

The heritage of accounting systems becomes evident in the specification of events. An event follows the most fundamental accounting principle (namely, the *duality principle*), which prescribes that every event can be broken down into atomic events, which involve pairs of agencies that have clearly identified roles as *givers* and *takers*. Atomic events, in turn, always occur in pairs involving the same agencies, such that the giver in one event of the pair is the taker in the other event of the pair, and vice versa.

For example, an event can be the acquisition of some good. The buyer is a giver of cash and a taker of the good to be bought, and the good provider is a taker of cash and a giver of the same good. An event can also be the utilization of the capability of an agency to execute a task. The provider of the capability is the giver of the actions to execute the task and the taker of whatever sort of remuneration is offered for the task, and the consumer of the capability is the taker of the execution of the task and the giver of the compensation for the execution of the task.

Agencies can be considered at different levels of detail. The corresponding events and exchange of resources that link agencies and their components must also be considered at the appropriate levels of detail, depending on what agencies are taken into account in the description of a business model.

The E-Commerce Integration Meta-Framework Project (http://www.ecimf.org) is a large project maintained by several institutions, such as the Royal Institute of Technology (Sweden), British Telecom, Hewlett-Packard, and the European consortium RosettaNet. Among its many outputs, we find the implementation of the REA artificial ontology using the *Protégé* editor (http://protege.stanford.edu).

The *Protégé* editor, Protégé-2000, has been developed by Stanford Medical Informatics at the Stanford University School of Medicine, with support from the National Library of Medicine, the National Science Foundation and the Defense Advanced Research Projects Agency. Protégé-2000 is available as free software under the open-source Mozilla Public License.

As presented in Noy et al. (2000, emphasis theirs):

> *the knowledge model of Protégé-2000 is frame-based: frames are the principal building blocks of a knowledge base. A Protégé-2000 ontology consists of classes, slots, facets, and axioms.* Classes are concepts in the domain of discourse. Slots *describe properties or attributes of classes.* Facets *describe properties of slots.* Axioms *specify additional constraints. A Protégé-2000 knowledge base*

4.3 ILLUSTRATIVE EXAMPLE I

includes the ontology and individual instances of classes with specific values for slots.

Classes in Protégé-2000 constitute a taxonomic hierarchy that admits multiple inheritance. The class hierarchy of Protégé-2000 is a directed acyclic graph. Classes are the nodes in the graph, and slots are what we have called attributes in our conceptual view of artificial ontologies. Protégé-2000 facets and axioms are the means in that editor to encode equations that describe the interdependences among slots and their values. The root element ⊤ is named :THING in Protégé-2000.

Each node in the acyclic graph is characterized by a collection of attributes. Using the Protégé-2000 jargon, each class is characterized by a collection of slots. Slots are created independently of classes and then linked to classes as desired.

An artificial ontology in Protégé-2000 admits specific *instances*, which are independent "knowledge bases" built in accordance with the encoded artificial ontology.

Artificial ontologies and their instances created using *Protégé* can be presented in several ways, amenable to browsing by human and artificial agents, such as database tables and HTML linked files.

The software *Protégé* and the corresponding documentation, can be found at http://protege.stanford.edu. It is extensible, and many plug-ins are available, including:

- Tools to export artificial ontologies and "knowledge bases" to several useful formats, such as Java files compatible with the JADE (Java Agent Development) framework.[4]

---

[4] JADE is a software framework for the implementation of multi-agent systems that can be distributed across a heterogeneous computer network. JADE is free software and is distributed by TILAB (http://www.telecomitalialab.com), the copyright holder, in open source software under the terms of the LGPL (Lesser General Public License Version 2).

- Visualization tools for graphical navigation along artificial ontologies and their instances.

- Tools for the management of multiple artificial ontologies, as described in Noy and Musen (2000). Essentially, this is done using tools that identify topological and literal similarities between artificial ontologies and support the *manual* alignment and merging of artificial ontologies. Two artificial ontologies are *aligned* when equivalent classes are identified between them. Two artificial ontologies can also be *merged* as a single artificial ontology based on the same principles that are used to align artificial ontologies.

The initial screen when the user activates Protégé-2000 is shown in Figure 4.1. At the top of this screen the user can see many folders. The first four are named Classes, Slots, Forms and Instances, and the others vary depending on which plug-ins are activated.

In Figure 4.1 we see the Classes folder. The leftmost region of the screen shows a representation of the class hierarchy. The rightmost region of the screen shows, for each selected class, its details, such as slots and constraints.

In Figure 4.2 we see the Classes folder for the REA artificial ontology, showing all classes used in this ontology. Classes in Protégé-2000 are either abstract or concrete. Abstract classes do not take direct instances, whereas concrete classes do.

The three upper classes in the REA artificial ontology are abstract.[5] They are called *ExchangeElement*, *ScriptElement* and *RecipeElement*. ExchangeElement is the class of concepts in the original REA model as proposed in McCarthy (1982). Every concept in the original REA model is related to some sort of *economic exchange*, which is an event involving some transfer of control of resources between agencies. ScriptElement is a class

---
[5] Abstract classes receive a small green *A* next to their names.

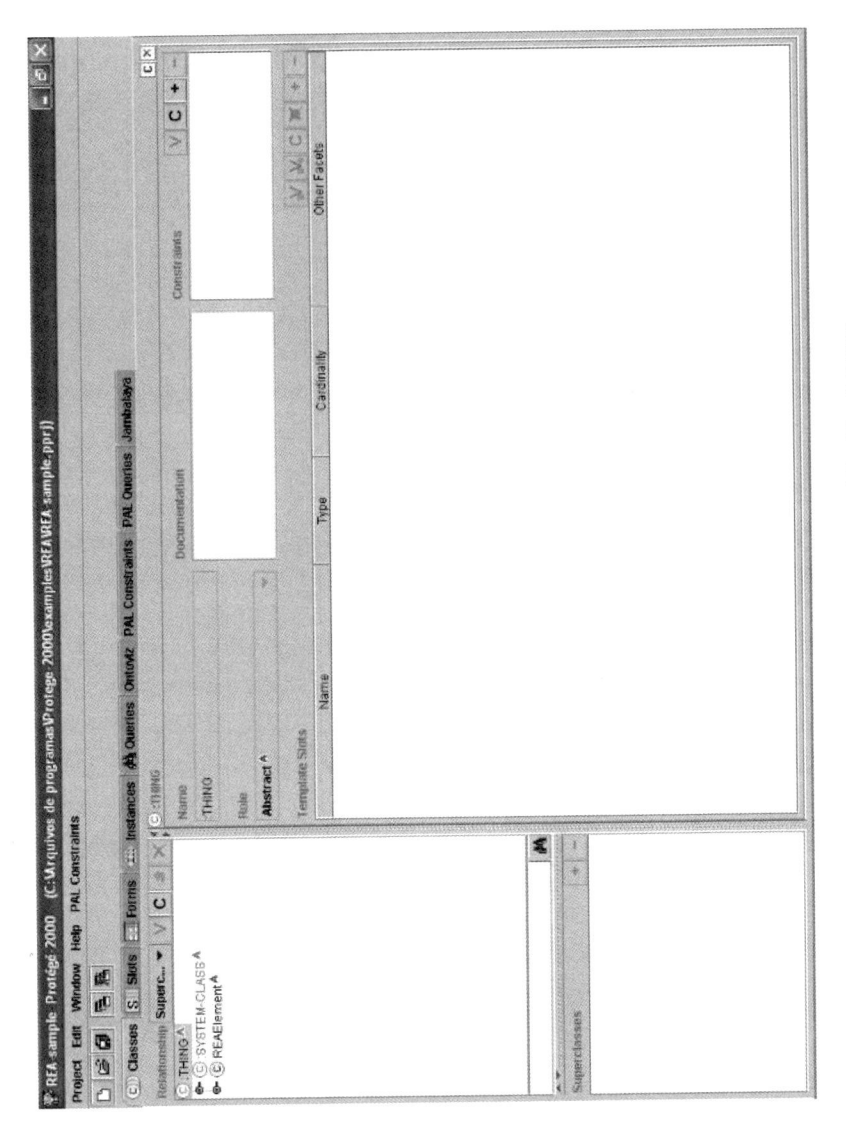

Figure 4.1 The initial screen of Protégé-2000.

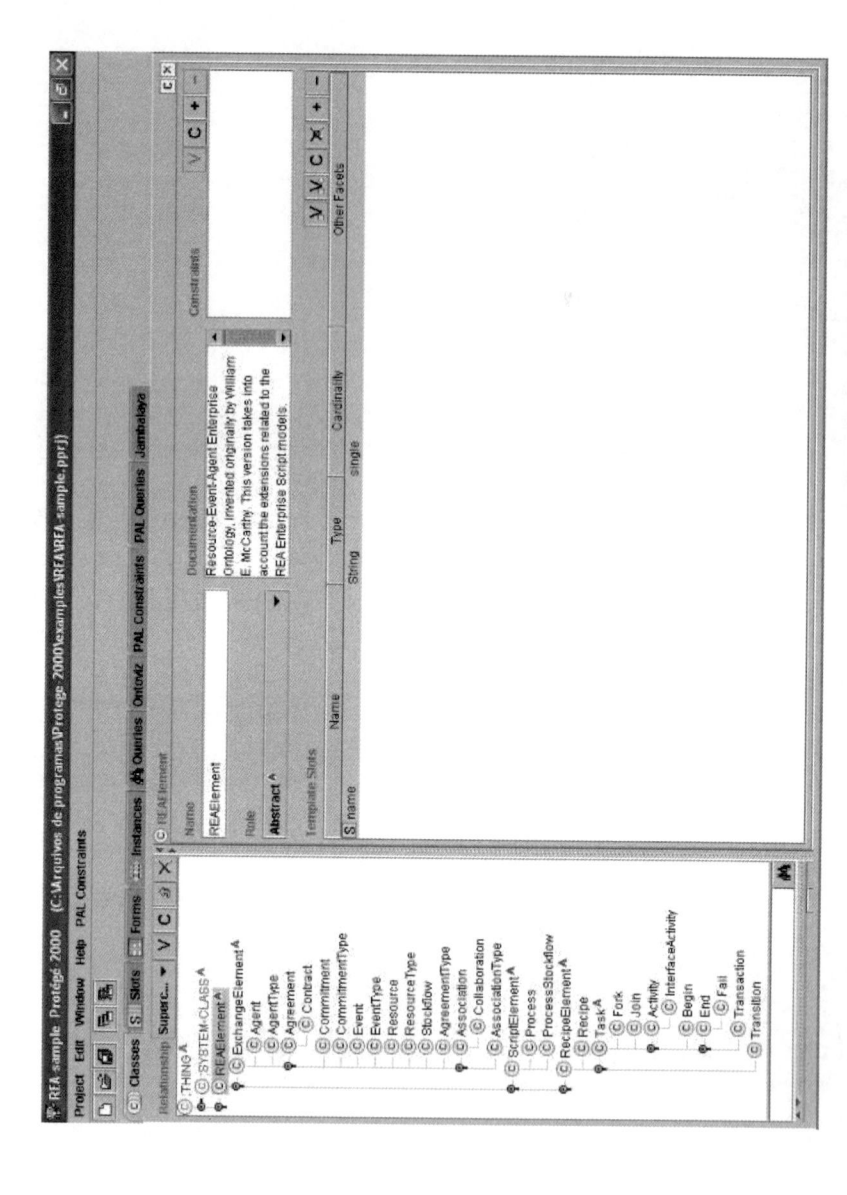

Figure 4.2 The classes in the REA artificial ontology.

of scripts, which are chains of economic exchanges. Recipe-Element is a class of recipes, which are collaboration protocols for fulfilling economic exchange events through a series of ordered tasks (activities).

The main concepts directly related to economic exchanges are, adopting the terminology found in the REA artificial ontology,[6] agents, events and resources. Hence, we find under the class ExchangeElement the classes Agent, Event and Resource. We also find the classes Commitment, StockFlow, Association and Agreement:

- An agent is characterized by a name, a set of associations, a set of events, a set of commitments and a set of resources. The resources are those under control of the agent, the events are those in which the agent takes part, the commitments are obligations to which the agent must engage and the associations characterize relationships between the agent and other agents.

- An event is characterized by a name, the corresponding transfer of control of resources and possibly by a dual event. For example, *buy* can be an event, with *sell* as a dual event.

- A resource is characterized essentially by a name. The model implemented by the E-Commerce Integration Meta-Framework Project (and reproduced here) adds the possibility of identifying whole–parts relationships among resources.

- Commitments characterize mutually agreed obligations among agents. A commitment is characterized by a name, an event (which is the event the agent is committed to execute), a resource that is transferred from the agent to another agent as

---
[6] Agents in the terminology found in the REA artificial ontology correspond to our concept of *agencies*.

a result of the event and *necessarily* a reciprocal commitment by another agent.

- StockFlow is the actual transfer of control of resources. It is characterized by a name, the name of the resource being transferred and the procedure that implements the transfer. The procedures taken into account in the REA model are give, take, produce, use and consume.

- Association is the means we find in the REA model to form agencies out of agencies. An association is characterized by a name (the name of an agency) and a set of components (which are REA agents).

- An agreement is a statement of collaboration among agents. It is characterized by a set of interlinked commitments.

There are other auxiliary classes in the REA model implementation proposed by the E-Commerce Integration Meta-Framework Project under the class ExchangeElement. The interested reader should consult the documentation found in http://www.ecimf.org.

An example of "knowledge base" is also presented together with the REA artificial ontology proposed by the E-Commerce Integration Meta-Framework Project. It is an instance of the REA artificial ontology for the business processes involved in renting a car.

In Figure 4.3 we have the Instances folder showing again the classes that comprise the ExchangeElement abstract class. Each concrete class is followed by a number, which indicates the number of instances that have been implemented for that class in the specific car rental "knowledge base".

For example, the class Agent has three instances, named Cashier, Customer and RentalAgent, as shown in Figure 4.3.

At the right side of the screen (Figure 4.3) we see the values of

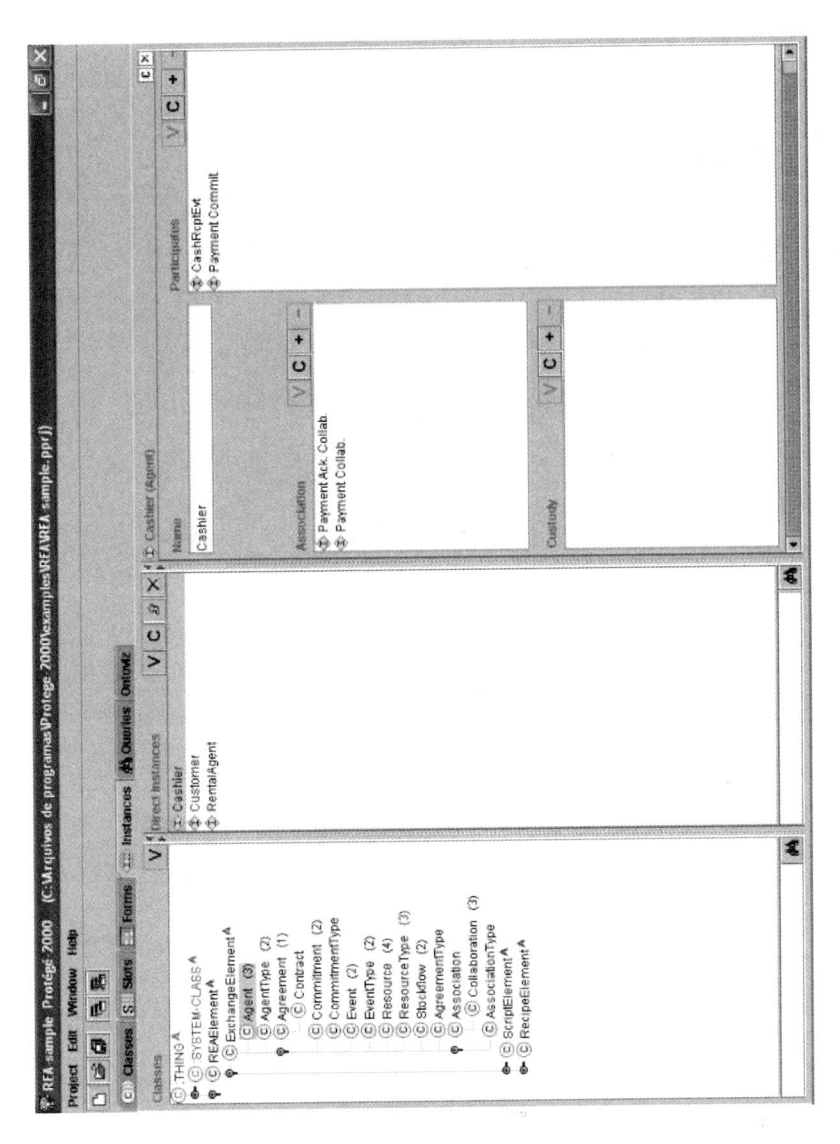

Figure 4.3 A car rental instantiation of the REA artificial ontology.

the slots associated with the class Agent, in the specific case of its instantiation as a cashier. As remarked above, an agent is characterized by its name, the associations (i.e., agencies) to which it belongs, the events in which it has a participating role, the commitments in which it is engaged and the resources it holds under control.

In a car rental situation, the events of interest are car rental and the corresponding payment. These two events must always come in pairs, hence they are called dual to each other. A cashier has direct participation in payment events, hence we find an event called CashRcpEvt (i.e., cash receipt event, or payment) as a slot for the cashier.

We have three sorts of associations in the car rental setting, which are called payment acknowledgement, payment and rental. Payment involves a customer and the cashier, payment acknowledgement involves the cashier and the rental agent, and rental involves the rental agent and the customer.

We also have two sorts of commitments: the commitment to pay for the rental of a car and the commitment to deliver a car that has been rented. The first one is a commitment between the customer and the cashier, and the second one is a commitment between the rental agent and the customer. Evidently, these two commitments are reciprocal of each other. The payment commitment triggers the cash receipt event, which has, as a stock flow action, the action *give cash*, whose effect is the transfer of control of money – a resource – from the customer to the cashier.

The car delivery commitment triggers the car rental event, which has, as a stock flow action, the action *take car*, whose effect is the transfer of control of a car – another resource – from the rental agent to the customer.

A cashier is thus characterized as belonging to payment and payment acknowledgement (two associations), and as participating in the payment commitment and payment events.

A customer is characterized as belonging to payment and rental associations, and as participating in the following

4.3 ILLUSTRATIVE EXAMPLE I

commitments and events: payment commitment, car delivery commitment, car rental event, and cash receipt event. A customer holds an amount of money that will be used to rent a car.

A rental agent is characterized as belonging to payment acknowledgement and rental associations, and as participating in the car delivery commitment and car rental event. A rental agent holds a car that will be delivered when it is rented.

The two basic resources we have are cash and cars. In the example presented by the E-Commerce Integration Meta-Framework Project, cars are organized in a hierarchy of types. For example, we have cars, trailers and Ford trailers. Every Ford trailer is a trailer, and every trailer is a car.

Finally, we have the definition of a rental agreement, which is a connection between a payment commitment and a car delivery commitment.

The diagram in Figure 4.4 summarizes the aspects of the REA artificial ontology described here. This diagram is based on the documentation of the REA artificial ontology delivered by the E-Commerce Integration Meta-Framework Project. Boxes are classes and arrows represent relationships between classes.

The diagram in Figure 4.5 shows the relationships between some specific instances of the classes of the REA artificial ontology, illustrating the intended meaning of those relationships.

The REA model is deservedly acknowledged as a fine piece of work. It extends the well-founded and universally accepted accounting model in a sound way, furnishing that model with the capability to keep track of economic events in real time.

For accountants, the information that matters is what can be written down in documents. Therefore, codification-based knowledge management is appropriate for those agencies.

Nevertheless, the REA model is certainly less known and less used than its proponents would expect. We believe this is not an isolated misfortune: the REA model was based on the assumptions that it would be *the most* appropriate model to describe the processes it takes into account and that therefore agencies who

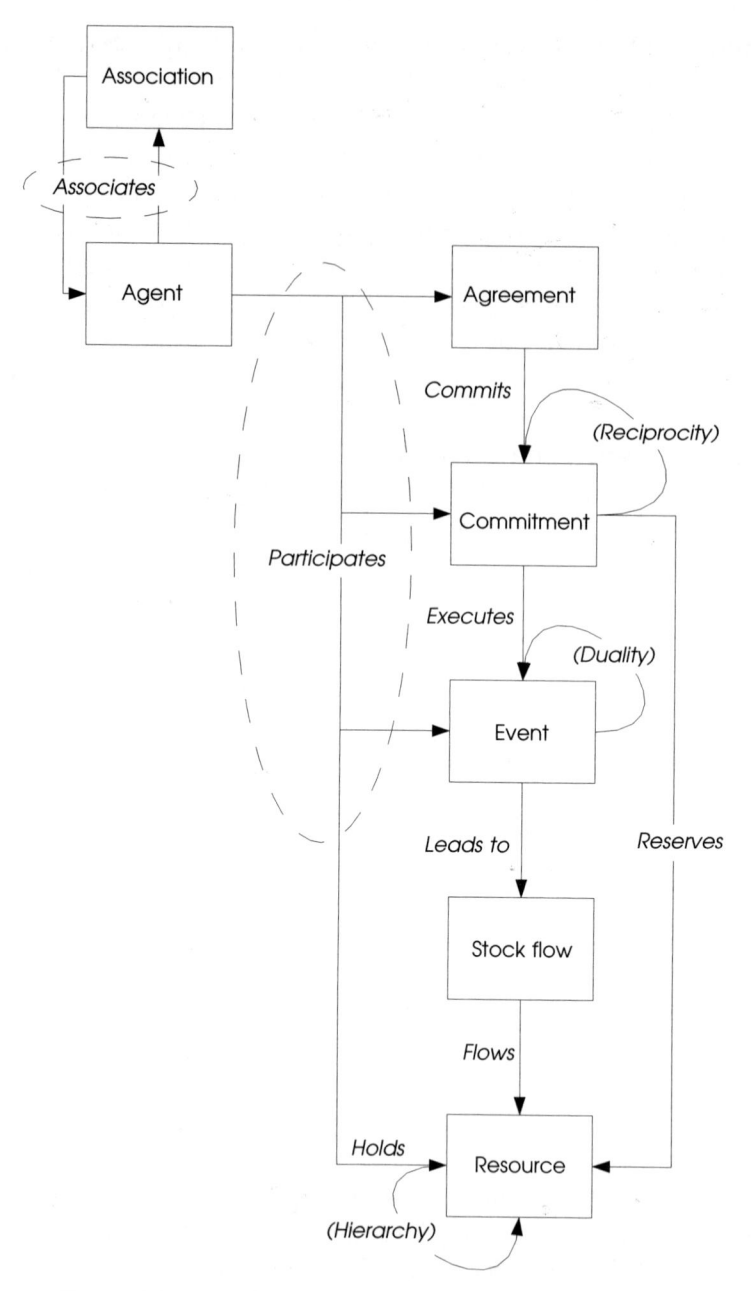

Figure 4.4 The REA artificial ontology – a simplified view.

4.3 ILLUSTRATIVE EXAMPLE I

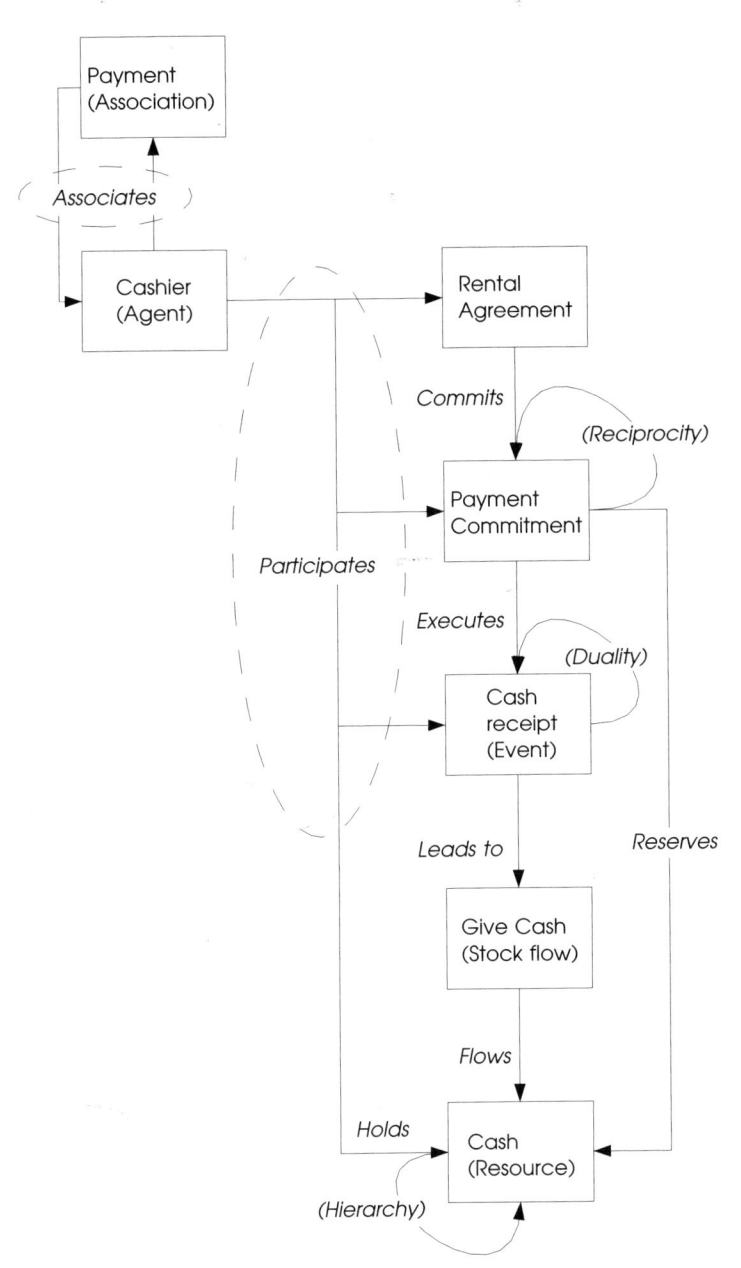

Figure 4.5 The REA artificial ontology – excerpts from the car rental instantiation.

perform economic transactions would be willing to give up their ways to structure, process and share information and to adopt what is proposed in the model.

This is unlikely to occur in large scale, even if the REA model is proposed by some respectful standards organization like the ISO. It may occur in small groups of agencies, though, who need to design their communication patterns in order to interact. For example, the car rental example described here can be appropriate for a single company or a small (e.g., regional) group of companies, although it may not necessarily be adequate as a universal standard to be forced onto every existing car rental agency in the world.

4.4 ILLUSTRATIVE EXAMPLE II: THE NATIONAL ACADEMIC CVs DATABASE IN BRAZIL – LATTES

A curriculum vitae is the organization of the academic and professional history of an individual. This organization is based on a *terminology*. Clearly, different terminologies can be adopted to organize a CV, depending on who is going to read it (the same individual may prepare rather different CVs, depending, e.g., on whether that individual seeks a position at a research university or a private company).

If we consider this terminology as the conceptualization of a person (personal data, academic background, professional skills and experience, and so on), we can well call an explicit and formal specification of this conceptualization an artificial ontology.

Since 1999 the Brazilian Ministry of Science and Technology, and the related funding agencies for research and higher education, have adopted a standardized format for the CVs of active researchers in Brazil. Although there is no explicit mention of the term "ontology", the first step to put this system to work has been an effort to convince the scientific community to embrace a proposed artificial ontology for CVs.

Essentially, the Lattes system[7] requires a user to download a collection of software products, with which a CV can be prepared. This CV resides at the user's computer, and it can also be uploaded to a central database in Brasilia. At the time of writing, 266,101 CVs had been included into the CV database in Brasilia.[8]

The CV is organized and encoded based on a collection of XML tags. These tags encode the artificial ontology adopted to build CVs. Based on this artificial ontology, many other software products have been developed and distributed by the Ministry of Science and Technology in Brazil, making room for a variety of searches that have helped to characterize the research community in Brazil. For example, researchers can be organized by research institution, by region or by subject, and an accurate mapping of scientific production in Brazil can be easily generated.

The Lattes CV system is a highly successful and appraised effort to organize *data* around a standardized terminology, to facilitate the utilization of these data. It is becoming an international standard and is in course of adoption by six more countries: Chile, Mexico, Venezuela, Colombia, Cuba and Portugal.

There are no indications that the Lattes CV system has explicitly employed the concepts, tools and systems to build and maintain artificial ontologies. Apparently, the terminology adopted and encoded as XML tags to build CVs was *hardcoded* as a collection of tags. In the following paragraphs we

---

[7] The system is named after Cesare Mansueto Giulio Lattes, the physicist who discovered the π-meson.
[8] The CV of Flávio Soares Corrêa da Silva belongs to this database. It can be reached at `http://genos.cnpq.br:12010/dwlattes/owa/prc_imp_ cv_int?f.cod=K4781644J4`.

are going to run a small experiment, namely the reconstruction of the Lattes CV artificial ontology[9] employing the editor Protégé-2000.

A simplified description of a Lattes CV is presented in Figure 4.6.

Each node in the graph is characterized by a collection of attributes. For example, the class Languages can be characterized by a slot *Language Name*, used to determine a specific language, and slots *Read, Write, Understand Spoken* and *Speak*, used to determine the skills of an individual with respect to specific languages. *Language Name* can be of type string, and *Read, Write, Understand Spoken* and *Speak* can admit a limited set of alternatives as values, as for example *Good, Average* and *Poor*.

The schema presented in Figure 4.6 was implemented as an artificial ontology in Protégé-2000 resulting in the class hierarchy given in Figure 4.7. Here, (c) identifies class names and (s) identifies slot names. A class labelled with a red (M) has multiple parents. In this figure, we highlight the class *Research Lines*, which is characterized by slots *Research Name* and *Research Goal*. *Research Name* is a single required slot (i.e., each instance of this class must have one and only one name). *Research Goal* is a multiple non-required slot (i.e., an instance of *Research Lines* can have zero, one or more than one declared research goals).

Slots are created independently of classes and then linked to each class as desired. In Figure 4.8 we have a partial view of the list of slots created for the Lattes CV.

Specific *instances* of this schema can be used to assemble individual CVs. In Figure 4.9 we show the process of creating an instance for the class *Personal Data*.

Finally, artificial ontologies and their instances created using Protégé-2000 can be presented in several ways, amenable to browsing by human and artificial agencies. In Figures 4.10 and 4.11, for example, we show the HTML files generated by

---

[9] Or, rather, a simplified version of the Lattes CV artificial ontology.

4.4 ILLUSTRATIVE EXAMPLE II

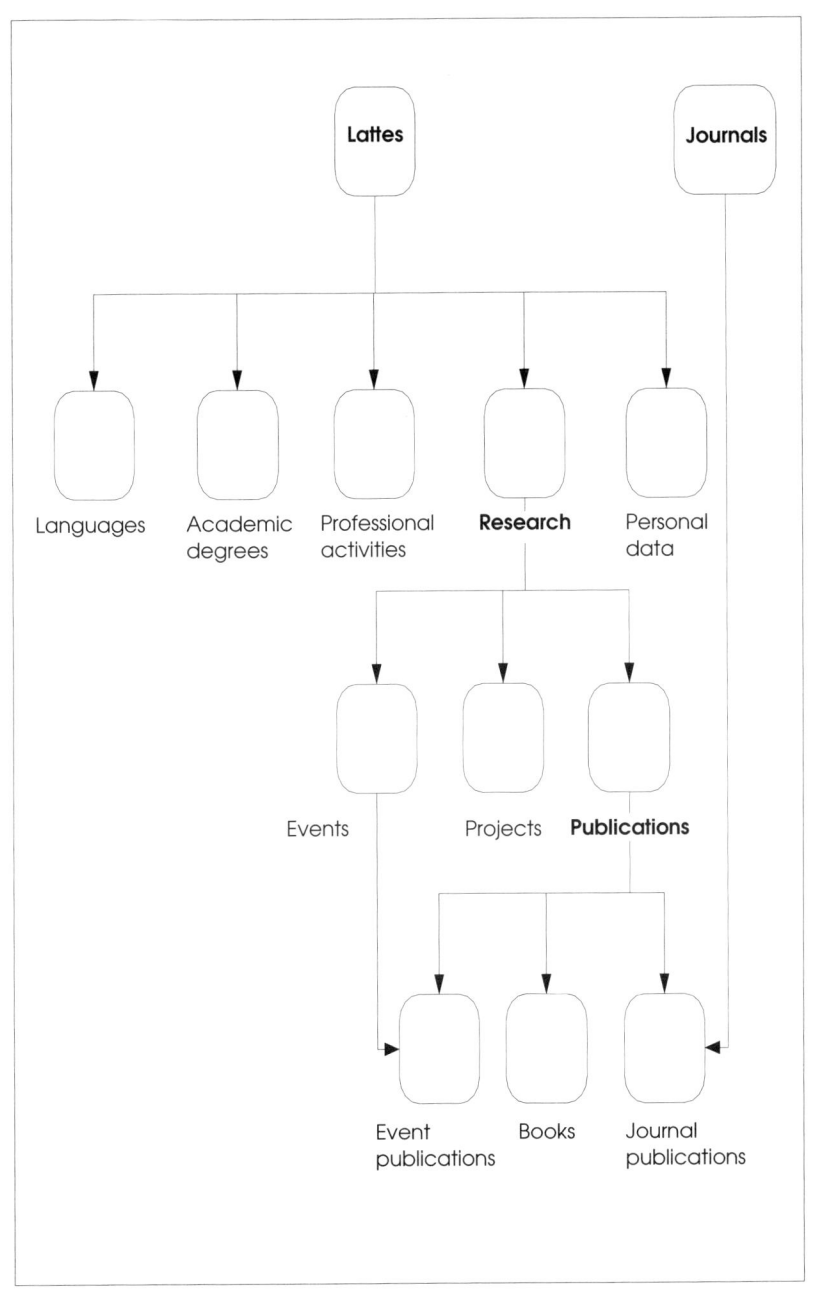

Figure 4.6 A class hierarchy for a CV.

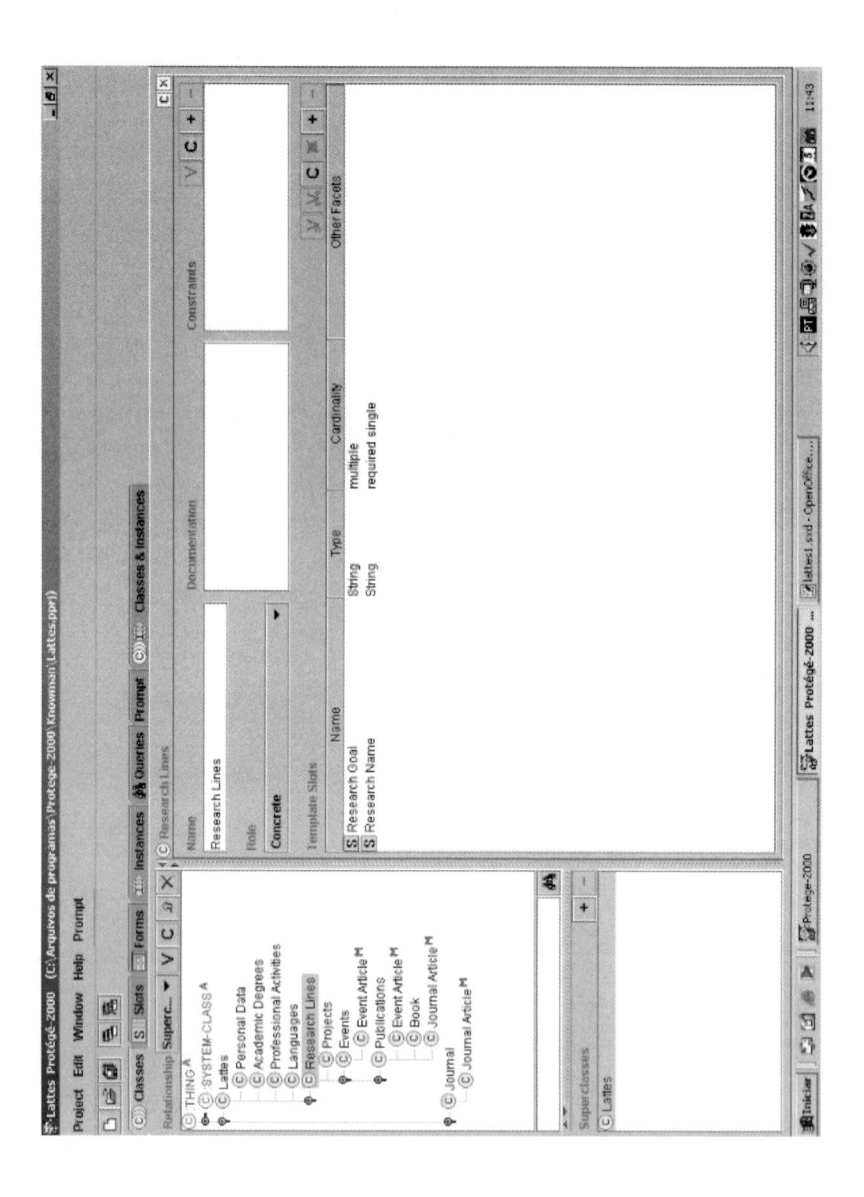

Figure 4.7 A class hierarchy for a CV in Protégé-2000.

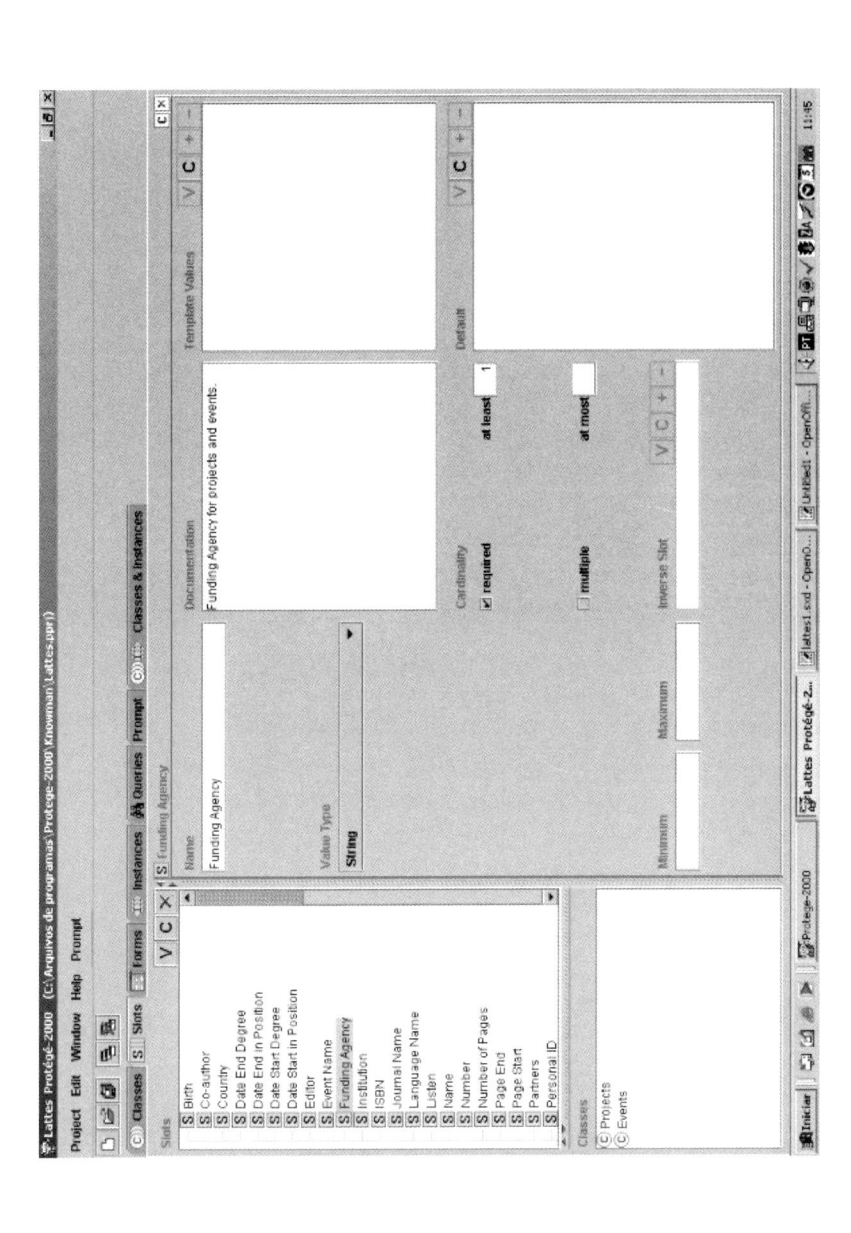

Figure 4.8 Slots for CVs in Protégé-2000.

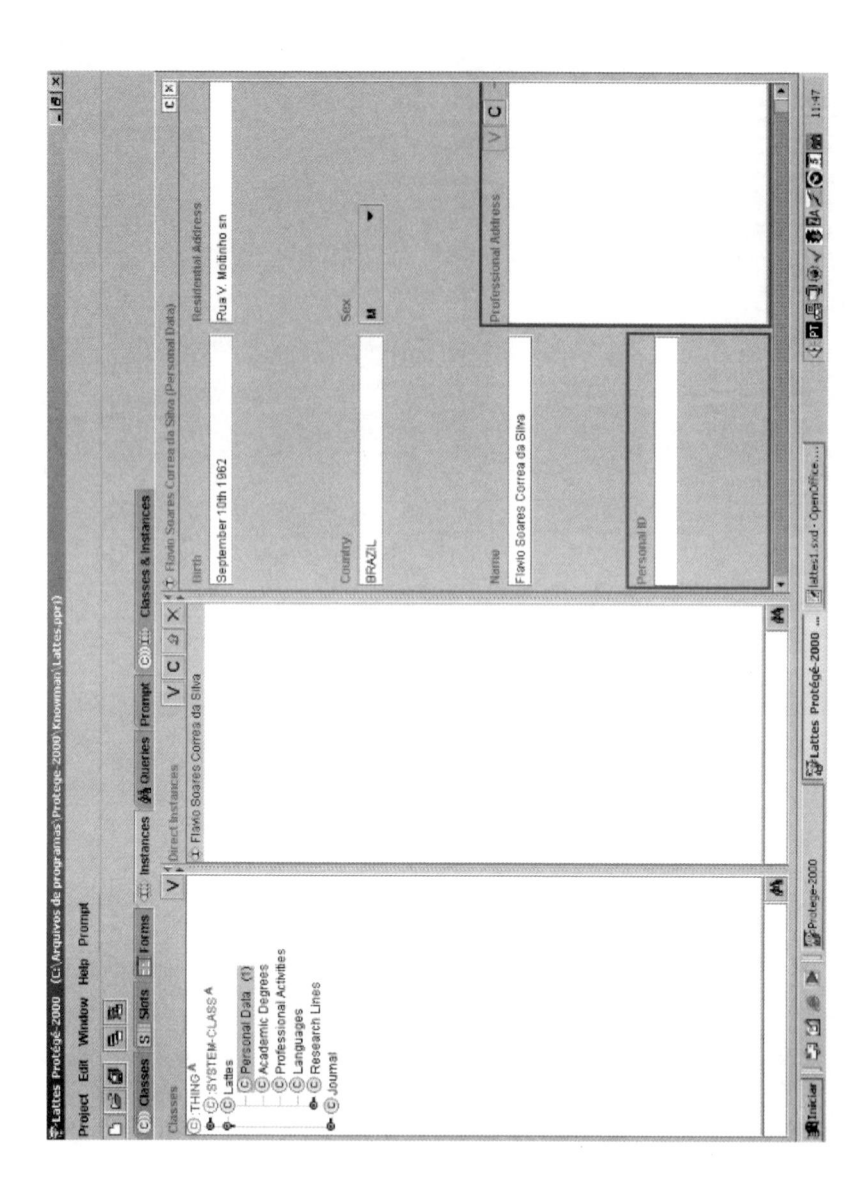

Figure 4.9 Creating an instance in Protégé-2000.

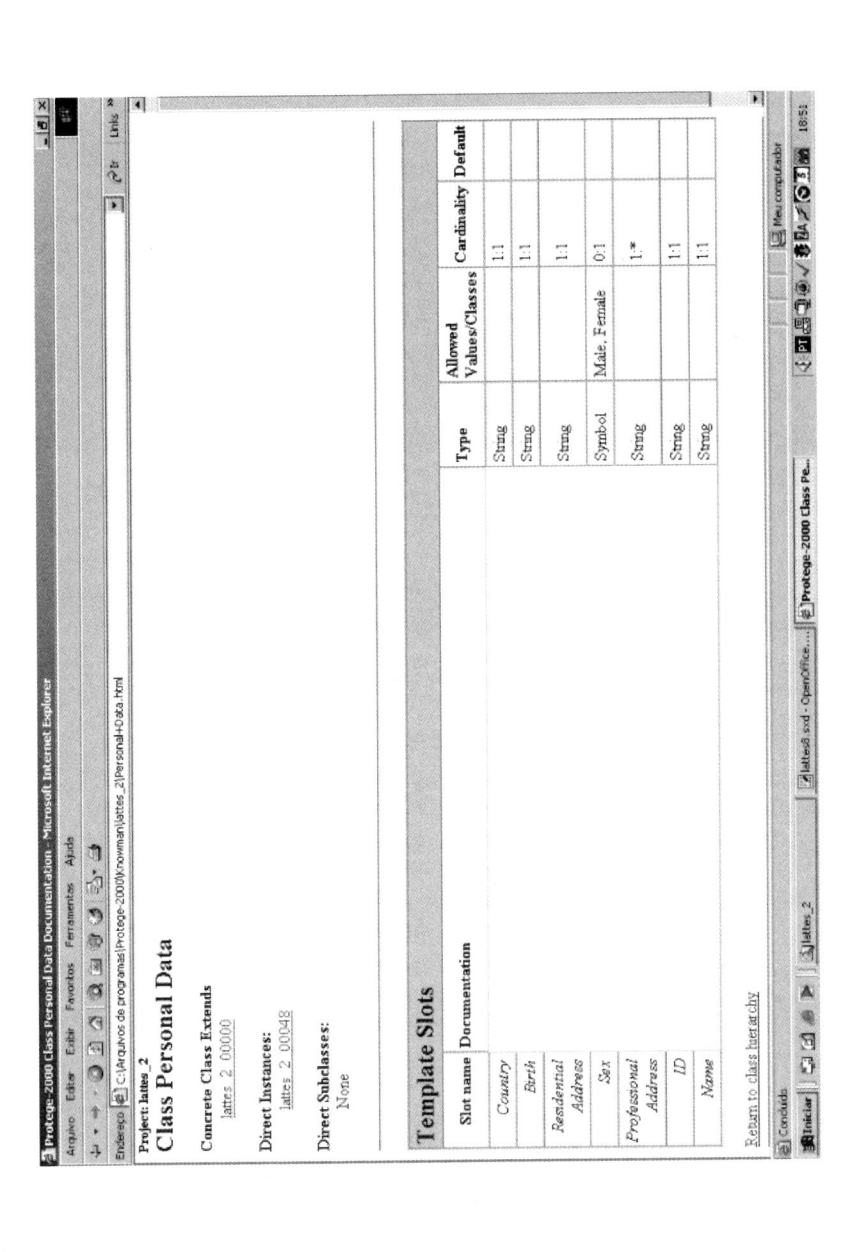

Figure 4.10 HTML presentation of a class generated by Protégé-2000.

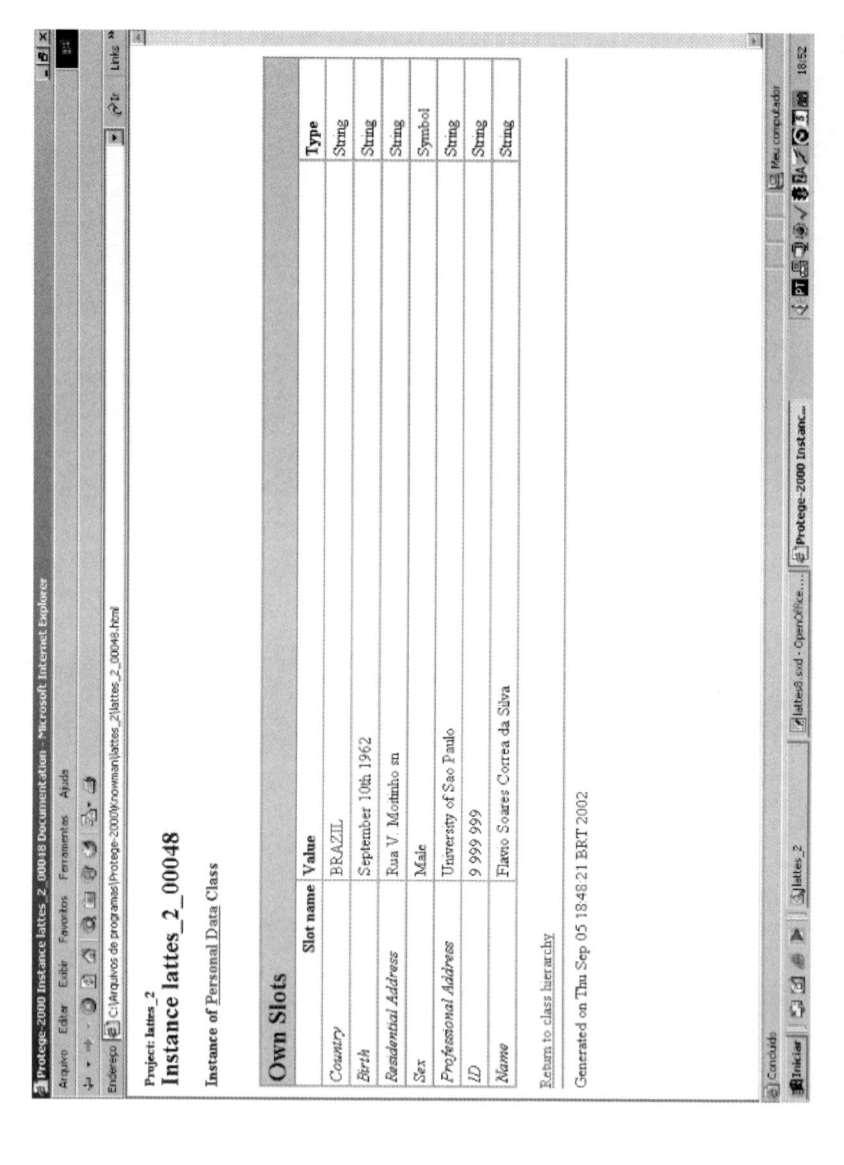

Figure 4.11 HTML presentation of the instance of a class generated by Protégé-2000.

Protégé-2000, respectively, for a class and its instance in the Lattes CV artificial ontology.

Assume that all we want to know about an individual can be written in that individual's CV. So, browsing through the artificial ontologies of CVs to identify the different representations of concepts for which we are looking, and then searching for appropriate instances of these concepts, could be regarded as a procedure for *knowledge sharing*. Indeed, this is the rather common procedure followed by most researchers nowadays. If the class *Publications* actually points to online versions of the published material of individuals, then each instance of this class presents the actual publications prepared by different authors. In a very specific sense, whenever we navigate through the Web and retrieve articles prepared by different authors, without ever contacting those authors for clarification, we accede to the view that all relevant information about those authors is expressed in the documents they have created.

Artificial ontologies can give access to otherwise inaccessible information about certain agencies, as happens with the academic use of the WWW to retrieve scientific articles published worldwide. On the other hand, this technology should *never* be considered as an alternative to direct contact between agencies (either artificial or human), whenever this contact is feasible. Knowledge coordination and management based on codification principles – and thus on artificial ontologies –*necessarily* impoverish and constrain the possibilities for knowledge sharing.

4.5 NATURAL ONTOLOGIES AND KNOWLEDGE COORDINATION

Artificial ontologies are an interesting way to organize information that can be represented symbolically, to facilitate the mapping of representations of information between systems.

Contrary to the initial claims regarding the power of artificial ontologies, however, they seem to be most useful to identify mappings among *small groups* of systems (e.g., mappings between pairs of systems).

Information that admits symbolic representation constitutes a fraction of the relevant information we find in agencies. Thus, artificial ontologies are deemed to capture only partial views of the world: clearly, the perception a customer has about the procedures and agencies involved in a car rental goes much beyond what can be expressed in the events above.

Large, centralized ontologies are attractive to managers because they promise to bring the control of the organization back to what was possible under classical management techniques. The problem is that they may also bring back the rigidity of agencies organized under the classical management tenets.

Furthermore, it can be difficult to convince all components of an organization to document their knowledge in terms of standardized artificial ontologies, even when it is possible. As occurred with the classical school of management, artificial ontologies – and the corresponding codification-based knowledge management techniques – have a tendency to overlook the relevance of knowledge that is not represented symbolically.

A more flexible way to employ artificial ontologies is as tools for *personal* organization of knowledge. Under this perspective, artificial ontologies are not necessarily to be shared, and the reason to use them is to ensure that every agency has an organized and well-defined method to carry on its tasks – which may not necessarily be broadcast to the whole organization. It becomes *common knowledge* (in the sense of Fagin et al., 1995) that every component of the organization is methodical in the same sense. However, the *content* and the *structure* of the ontology of each agency, as well as the *content* and the *structure* of the knowledge of each agency, are not common knowledge.

Clearly, this utilization of artificial ontologies (as tools to organize individual knowledge instead of tools to share knowledge) does not preclude the more common one outlined above. Two independent artificial ontologies may coexist within an agency: the *internal* artificial ontology that is used to organize its methodology to work and the *external* artificial ontology that is used to ensure that the agency can communicate effectively with other agencies. None of these information structures, however, captures the *real* (or "natural") ontology of that agency, which is the whole set of concepts whose existence is acknowledged by the agency as a whole – including those that defy symbolic representation.

In the following chapter we present a concrete, proposed application of internal artificial ontologies, called *structures of capability providers*. These structures have been designed to permit the management of communities of practice, a central concept for personalization-based knowledge management.

We describe this conceptual tool in detail and show how it can be used for knowledge management. Indeed, structures of capability providers are a non-intrusive way to ensure knowledge coordination and to make good use of the full knowledge – symbolically represented or otherwise – of agencies. They can be used for agencies based on human agents as well as software agents, as we also show.

REFERENCES

Corrêa da Silva, F. S. and Meneses, E. X. (2001) "Integração de agentes de informação," in A. T. Martins and D. L. Borges (eds), *Anais do XXI Congresso da Sociedade Brasileira de Computação* (Vol. 3) – Jornada de Atualização em Inteligencia Artificial, Brazil [in Portuguese].

Corrêa da Silva, F. S. and Meneses, E. X. (2002) "Expressing systems capabilities for knowledge coordination," in *Proceedings of AAMAS – Autonomous Agents and Multi-agent Systems 2002, Italy*.

Corrêa da Silva, F. S., Vasconcelos, W. W., Agustí-Cullell, J., Robertson, D. S. and Melo, A. C. V. (1999) "Why ontologies are not enough for knowledge sharing," in *12th International Conference on Industrial and Engineering Applications of Artificial Intelligence and Expert Systems, Egypt*.

Corrêa da Silva, F. S., Vasconcelos, W. W., Robertson, D. S., Brilhante, V., Melo, A. C. V., Finger, M. and Agustí-Cullell, J. (2001) "On the insufficiency of ontologies: Problems in knowledge sharing and alternative solutions," *Knowledge Based Systems*, **14**(7).

Davenport, T. H. and Prusak, L. (1998) *Working Knowledge*, HBS Press.

Ding, Y., Fensel, D., Klein, M. and Omelayenko, B. (2002) "The semantic web: Yet another hip?," *Data and Knowledge Engineering*, **41**, 205–227.

Fagin, R., Halpern, J. Y., Moses, Y. and Vardi, M. Y. (1995) *Reasoning About Knowledge*, MIT Press.

Fensel, D. (2001) *Ontologies: A Silver Bullet for Knowledge Management and Electronic Commerce*, Springer-Verlag.

Fensel, D., van Harmelen, F., Horrocks, I., McGuinness, D. L. and Patel-Schneider, P. F. (2001) "OIL: An ontology infrastructure for the semantic web," *IEEE Intelligent Systems*, **16**(2), March–April.

Garvin, D. A. (1993) "Building a learning organization," *Harvard Business Review*, July–August, 81–91.

Geerts, G. L. and McCarthy, W. E. (1999) "An accounting object infrastructure for knowledge-based enterprise models," *IEEE Intelligent Systems*, July/August, 1–6.

Gruber, T. R. (1993) "A translation approach to portable ontologies," *Knowledge Acquisition*, **5**(2), 199–220.

Hansen, M. T., Nohria, N. and Tierney, T. (1999) "What's your strategy for managing knowledge?," *Harvard Business Review*, **106**, March–April.

Haugen, R. and McCarthy, W. E. (2000) "REA, a semantic model for Internet supply chain collaboration," *OOPSLA – ACM Conference on Object-oriented Programming, Systems, Languages, and Applications –*

REFERENCES

Business Objects and Component Design and Implementation Workshop VI: Enterprise Application Integration, Minneapolis, MN.

Kalfoglou, Y., Menzies, T., Althoff, K. and Motta, E. (2000) "Metaknowledge in systems design: Panacea ... or undelivered promise?," *The Knowledge Engineering Review*, **15**(4), 381–404.

Kinny, D. (2001) "Reliable agent communication – a pragmatic perspective," *New Generation Computing*, **19**, 139–156.

Klusch, M. (2001) "Information agent technology for the Internet: A survey," *Data and Knowledge Engineering*, **36**, 337–372.

McCarthy, W. E. (1982) "The REA accounting model: A generalized framework for accounting systems in a shared data environment," *The Accounting Review*, 554–578.

McCarthy, W. E. and Geerts, G. L. (1997) "Modeling business enterprises as value-added process hierarchies with resource-event-agent object templates," in J. Sutherland and D. Patel (eds), *Business Object Design and Implementation*, Springer-Verlag, pp. 113–128.

Nonaka, I. (1991) "The knowledge creating company," *Harvard Business Review*, November–December, 96–104.

Noy, N. F. and Musen, M. A. (2000) *PROMPT: Algorithm and Tool for Automated Ontology Merging and Alignment*, AAAI/IAAI TX.

Noy, N. F., Fergerson, R. W. and Musen, M. A. (2000) "The knowledge model of Protégé-2000: Combining interoperability and flexibility," *2nd International Conference on Knowledge Engineering and Knowledge Management (EKAW 2000)*, Juan-les-Pins, France.

Senge, P. (1990) *The Fifth Discipline*, Currency/Doubleday.

Starkey, K. (ed.) (1992) *How Organizations Learn*, Thomson Business Press.

Uschold, M. (1998) "Where are the killer apps?," *Workshop on Applications of Ontologies and Problem Solving Methods, ECAI98*, Brighton, UK.

5

Capabilities

Reach what you cannot. – Nikos Kazantzakis

In this chapter we present a conceptual tool for management of the personal structured records of capability providers. This tool can be useful to stimulate and guide the formation of communities of practice, a key concept for personalization-based knowledge management.

The proposed tool is called *structure of capability providers*. As shown in the remainder of this chapter, it can be useful to discipline the flow of knowledge within an agency. For this reason, we call it a tool for *knowledge coordination*.

In brief, each agency must have a structure of capability providers – or the resources to build one dynamically whenever necessary – for each possible task presented to it. Structures of capability providers are built based on artificial ontologies of capabilities. These artificial ontologies, as well as the structures built using them, must not be shared among agencies. It is required, however, of every agency belonging to an organization to have and to use structures of capability providers to discipline the delegation of tasks within the organization. It must be

---

Knowledge Coordination   F. S. Corrêa da Silva and J. Agustí-Cullell
© 2003 John Wiley & Sons, Ltd   ISBN: 0-470-85832-X

common knowledge within the organization that every one of its components distributes its tasks based on disciplined and methodical rules, although these rules must not be communicated.

A structure of capability providers is a set of agencies, together with a set of partial orders of these agencies. The agencies are those known by the holder of the structure to be prepared to execute a given task – hence, they are "called capability providers". Each partial order captures some way to give preference to each agency in the set of capability providers (e.g., give preference to the fastest one, or to the most reliable one, or to the cheapest one).

The flow of knowledge within an organization is equivalent to the flow of capabilities in our setting. The management of the partial orders belonging to structures of capability providers is the procedure we propose to organize and control the flow of capabilities. If the components of an organization always try to delegate the tasks they undertake according to these partial orders, it is natural for them to have more frequent contact with some agencies than with others (namely, with those agencies that come first in their preferences expressed by the partial orders). Thus, given a task, an agency that undertakes it coupled with some criteria to select a partial order of capability providers, a hierarchy of agencies can be constructed. If we call the agencies that come first in a partial order *closer* to the holder of the corresponding structure of capability providers, we can see this structure as the documentation of a community of practice, as argued in the remainder of this chapter.

Knowledge coordination based on structures of capability providers can be *less intrusive* than knowledge coordination based on centralized artificial ontologies. A more subtle but important feature of this tool is that it does not limit management models to knowledge that can be symbolically represented, thus permitting incorporation of implicit (aka. *tacit*) *knowledge* explicitly in the system. On the other hand, a knowledge coordination system that uses our proposed tool requires the direct intervention of specific agents pertaining to the organization. Hence, such a system would carry the advantages and pitfalls of agencies based on personalization as opposed to agencies based on codification for knowledge management.

A capabilities-based knowledge coordination system is weaker in retaining knowledge than one based on explicit knowledge representation (e.g., by means of centralized artificial ontologies). It should be remarked, however, that a significant portion of the knowledge that has to be coordinated within an organization is not explicitly represented anyway, so the alternative of just identifying its sources and how they can be accessed can be of great service. Our proposal therefore shall be regarded as a complement to the *ontological engineering approach* advocated by, for example, Fensel (2001) and Fensel et al. (2001) and outlined in Chapter 4.

Our proposal is akin to the idea of *Knowledge Maps* (Davenport and Prusak, 1998; Liebowitz, 2001; Tiwana, 1999). Indeed, in chap. 4 of Davenport and Prusak (1998, "Knowledge codification and coordination") the desired features of a knowledge map are given: knowledge maps should be pointers to where the knowledge is, without actually representing knowledge itself; they should tell people where to go when needing some sort of knowledge.

In Liebowitz (2001) we find that the name "knowledge map" has been used to identify four types of diagrammatic identification of sources of knowledge:

1. The *a posteriori* mapping of actual communication that occurs among, for example, departments within an organization. This can be useful to identify unbalanced communication that may exist at the organization (e.g., when two departments that should have highly synchronized activities communicate less than expected).

2. The connection of capabilities with their potential providers.

3. The identification of capabilities that may be required within the organization and a diagnosis of their availability.

4. A taxonomy of capabilities.

Our proposal encompasses the last three (i.e., knowledge maps based on the concept of capabilities).

5.1 MANAGING CAPABILITIES

Our model is based on the concepts of *agency* and *task*. A task is a goal proposed to an agency. Once a task is executed, the value of a non-empty set of attributes is changed. Hence, a task is an action, and its execution requires knowledge: an agency must hold a specific capability to execute a task.

A task is represented as a relation between sets of states. We define a set of interchangeable states as the *input* set and another set of interchangeable states as the *output* set. We also define a set of constraints to express the interdependencies between attributes of input and output.

A task can be *refined* into subtasks. Refinement rules are not standardized, and each agency may have its own set of rules.

Just as an example, to show what can be meant by a set of refinement rules, we propose a simple set of rules that can be useful for a software agency. The rules we propose are:

- Sequential decomposition – a task t can be decomposed into two tasks t_1 and t_2, such that the input of t_2 is a subset of the union of the input and the output of t_1, the input of t_1 is the same as the input of t and the output of t_2 is the same as the output of t.

5.1 MANAGING CAPABILITIES

- Parallel decomposition – a task t can be decomposed into two tasks t_1 and t_2, such that the union of the inputs of t_1 and t_2 is the input of t, and the union of the outputs of t_1 and t_2 is the output of t.

- conditional decomposition – a task t can be decomposed into two tasks t_1 and t_2 plus a decision procedure δ. The union of the inputs of t_1, t_2 and δ is the input of t. Depending on the specific state in δ, either t_1 or t_2 (but not both) takes the task. Thus, the output of t is either the output of t_1 or the output of t_2.

- Iteration – a task t can be decomposed into two tasks t_1 and t_2 plus a decision procedure δ as above. The union of the inputs of t_1, t_2 and δ is the input of t. The output of t is the output of t_1, but the output of t_2 is a subset of the input of t. Depending on the specific state in δ, either t_1 takes the task and finishes it or t_2 takes the task, executes it and feeds back its output as a new input.

These decomposition rules are depicted in Figures 5.1–5.4.

The opposite of task refinement is task aggregation. Task aggregation dictates rules for composition of agencies to form a larger agency. For example, considering the refinement rules presented above, if agency Ag_i can execute task t_1 and agency Ag_j can execute task t_2, and if task t is the parallel aggregation of t_1 and t_2, then the agency required to undertake task t is

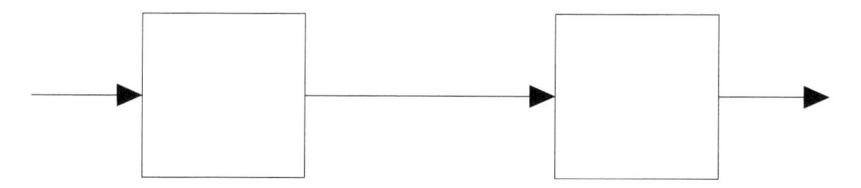

Figure 5.1 Sequential decomposition.

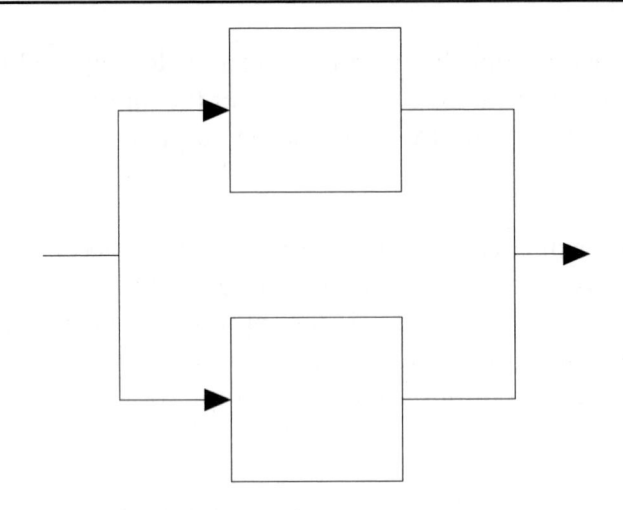

Figure 5.2 Parallel decomposition.

composed of Ag_i and Ag_j. In this specific example, parallel aggregation requires that the components of the larger agency are disjoint agencies. Sequential aggregation, conditional aggregation and iteration do not have this requirement.

The holder of a structure of capability providers has for

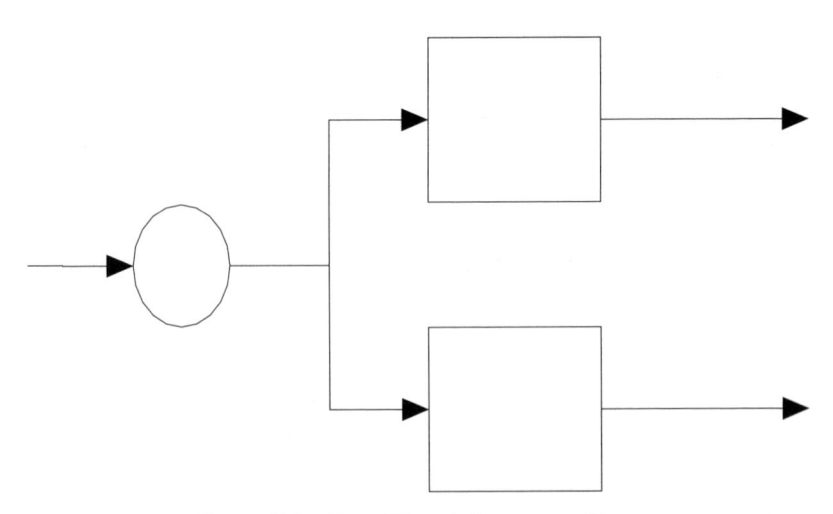

Figure 5.3 Conditional decomposition.

5.1 MANAGING CAPABILITIES 141

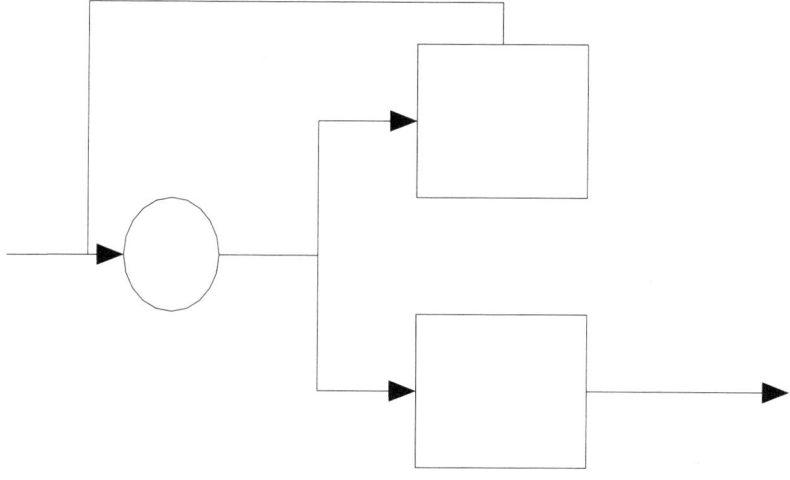

Figure 5.4 Iteration.

each incoming task a partially ordered set of agencies that can undertake that task. This set may or may not include the holder of the structure itself. The partial order determines the sequence in which agencies must be consulted about whether they would be willing to undertake that task. As proposed above, different partial orders can be used, depending on the appropriate criteria to be used to select a capability provider.

The procedure to select a capability provider is as follows: following the appropriate partial order, each capability provider is selected, one by one. For each capability provider, the agency that has received the task estimates an idealized quality of service and then proposes the task to the capability provider. Three possibilities can occur:

1. The capability provider refuses the task (it may be busy, out of service, etc.). In this case, the agency must propose the task to the next capability provider in the list.

2. The capability provider accepts the task, but commits itself to a real quality of service that is inferior to the idealized one

estimated by the agency that is proposing the task. In this case, the agency must decide whether it delegates the task to this capability provider or proposes it to the next provider in the list. This decision can be made, for example, based on a comparison between the real quality of service proposed by the present capability provider and the idealized quality of service estimated for the next capability provider in the list.

3. The capability provider accepts the task and commits itself to a real quality of service that is at least as good as the one estimated by the agency that is proposing the task. In this case, the agency must delegate the task to this capability provider and end the procedure.

This procedure is depicted in Figure 5.5. In Barnard (1938) we find what he calls "the executive functions":

- to provide the organization with a system of communication (among agents);

- to identify goals in the organization; and

- to guide the use of communication channels to facilitate the achievement of goals.

Knowledge coordination is performed using our proposed tool through this last item. Since we intend structures of capability providers to be accessed – and possibly updated – by knowledge coordinators, we propose that these structures be represented in a uniform way for all agencies. In what follows, we present a concrete proposal for the representation of these structures.

5.1 MANAGING CAPABILITIES

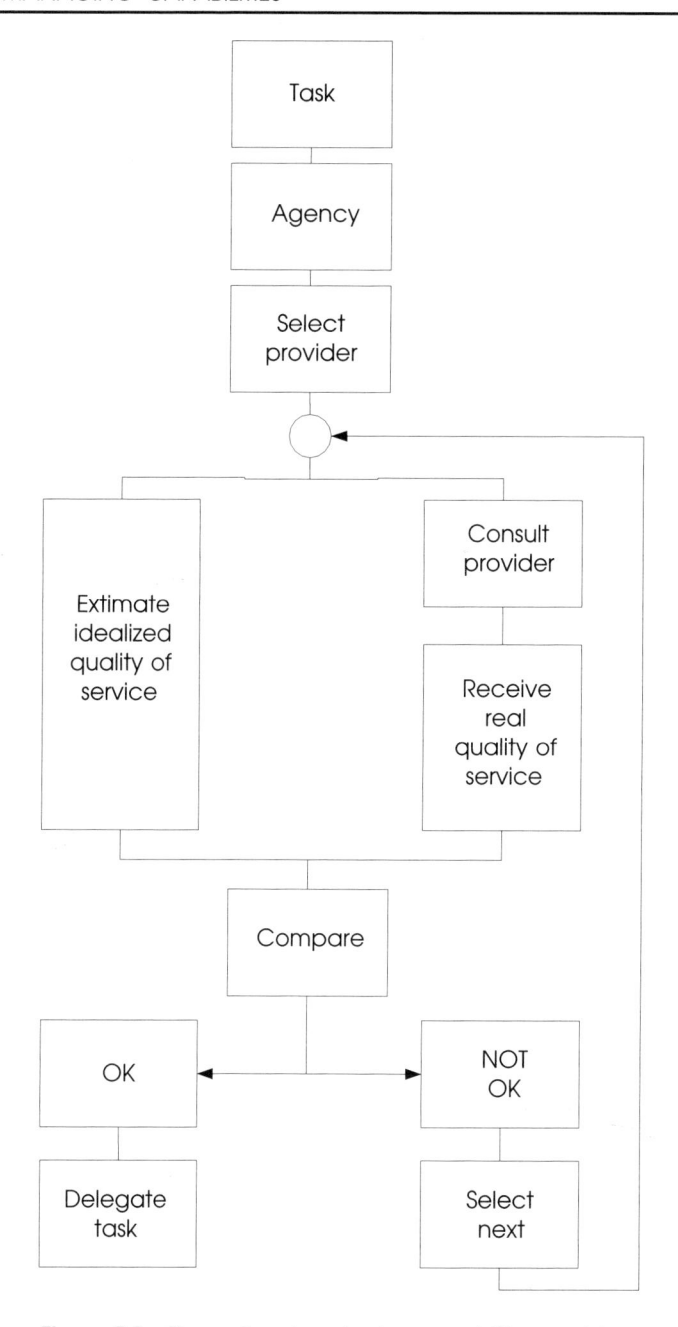

Figure 5.5 Procedure to select a capability provider.

5.2 STRUCTURES OF CAPABILITY PROVIDERS

Capabilities and tasks are interchangeable and represented the same way. Hence, a capability is a triple $(\mathcal{I}, \mathcal{O}, \mathcal{C})$, in which $\mathcal{I}$ is the input set of states, $\mathcal{O}$ is the output set of states and $\mathcal{C}$ is a set of constraints relating the values of the attributes of $\mathcal{I}$ and $\mathcal{O}$.

Capabilities can be refined in the same way as tasks. Each agency in an organization can have different representations for capabilities and tasks, and each agency can have different refinement rules for tasks and capabilities. Hence, for an agency to be able to delegate a task to another agency, it must be able to translate the task to the representation of the service provider. Once the task is executed, it must also be able to translate back the obtained output state from the representation of the provider to its own representation.

We start with the simple case of one agency requesting the execution of a task and only one agency as a candidate capability provider. Let $\mathcal{H}$ be the initial holder of the task and $\mathcal{P}$ be the candidate capability provider.

We assume that a task has been proposed to $\mathcal{H}$, who thus becomes the holder of that task. The agency $\mathcal{H}$ must have an internal representation of this task, denoted as $(\mathcal{I}_\mathcal{H}, \mathcal{O}_\mathcal{H}, \mathcal{C}_\mathcal{H})$.

The same task can have a different internal representation in the agency $\mathcal{P}$. We denote this representation as $(\mathcal{I}_\mathcal{P}, \mathcal{O}_\mathcal{P}, \mathcal{C}_\mathcal{P})$.

If agency $\mathcal{H}$ is interested in delegating this task to agency $\mathcal{P}$, then it must be able to translate the task from its own representation to the representation of $\mathcal{P}$ and vice versa. Let:

- $Tr(\mathcal{I}_\mathcal{H} \to \mathcal{I}_\mathcal{P})$ be a translation of the input set of states from the representation of $\mathcal{H}$ to the representation of $\mathcal{P}$;

- $Tr(\mathcal{O}_\mathcal{H} \to \mathcal{O}_\mathcal{P})$ be a translation of the output set of states from the representation of $\mathcal{H}$ to the representation of $\mathcal{P}$;

5.2 STRUCTURES OF CAPABILITY PROVIDERS

- $Tr(\mathcal{C}_\mathcal{H} \rightarrow \mathcal{C}_\mathcal{P})$ be a translation of the set of constraints between input and output states from the representation of $\mathcal{H}$ to the representation of $\mathcal{P}$; and

- $Tr(\mathcal{O}_\mathcal{H} \leftarrow \mathcal{O}_\mathcal{P})$ be a translation of the output set of states from the representation of $\mathcal{P}$ to the representation of $\mathcal{H}$.

The delegation of a task from $\mathcal{H}$ to $\mathcal{P}$ goes as follows:

1. agency $\mathcal{H}$ undertakes the task, which is internally represented as $(\mathcal{I}_\mathcal{H}, \mathcal{O}_\mathcal{H}, \mathcal{C}_\mathcal{H})$. A specific input value $i_\mathcal{H} \in \mathcal{I}_\mathcal{H}$ is also given, to which a specific output value $o_\mathcal{H} \in \mathcal{O}_\mathcal{H}$ is expected from $\mathcal{H}$;

2. agency $\mathcal{H}$ translates the task and the specific input value to the representation of $\mathcal{P}$, using the appropriate translations given above;

3. the task is proposed to agency $\mathcal{P}$;

4. assuming that agency $\mathcal{P}$ undertakes the task, a specific output value $o_\mathcal{P} \in \mathcal{O}_\mathcal{P}$ is produced;

5. using the appropriate translation given above, this output value is translated to $o_\mathcal{H} \in \mathcal{O}_\mathcal{H}$.

Now let $\mathcal{H}$ be as above and $\mathcal{P}_1, \ldots, \mathcal{P}_n$ be a set of alternative candidate providers of a capability. These providers must be comparable, and the most flexible way to ensure this is to make them partially ordered. A partial order $\subseteq$ is a relation between pairs of $\mathcal{P}_i$ with the following properties:

- $\mathcal{P}_i \subseteq \mathcal{P}_i$ for any $\mathcal{P}_i$ (reflexivity);

- if $\mathcal{P}_i \subseteq \mathcal{P}_j$ and $\mathcal{P}_j \subseteq \mathcal{P}_i$, then $\mathcal{P}_i$ and $\mathcal{P}_j$ are the same (antisymmetry); and

- if $\mathcal{P}_i \subseteq \mathcal{P}_j$ and $\mathcal{P}_j \subseteq \mathcal{P}_k$, then $\mathcal{P}_i \subseteq \mathcal{P}_k$ for any $\mathcal{P}_i, \mathcal{P}_j, \mathcal{P}_k$ (transitivity).

This is a rather technical way of ensuring that we have a coherent order of preference for $\mathcal{P}_1, \ldots, \mathcal{P}_n$ that does not necessarily require every pair of alternative providers to be comparable with each other. If $\mathcal{P}_i \subseteq \mathcal{P}_j$, then agency $\mathcal{H}$ shall try to delegate the task first to $\mathcal{P}_j$, and only to $\mathcal{P}_i$ if $\mathcal{P}_j$ refuses the task.

Each agency $\mathcal{P}_1, \ldots, \mathcal{P}_n$ has an estimated quality of service associated with it that agrees with the partial order; that is, if we denote the estimated quality of service of $\mathcal{P}_i$ as $q^e(\mathcal{P}_i)$, then, if $\mathcal{P}_i \subseteq \mathcal{P}_j$, we get $q^e(\mathcal{P}_i) \leq q^e(\mathcal{P}_j)$.

Finally, each agency $\mathcal{P}_1, \ldots, \mathcal{P}_n$ has a real quality of service associated with it, denoted as $q^r(\mathcal{P}_1), \ldots, q^r(\mathcal{P}_n)$. The estimated qualities of service are produced by $\mathcal{H}$, whereas each real quality of service is produced by the corresponding provider $\mathcal{P}_i$.

With all this in hand, the procedure presented in Figure 5.5 can be implemented. In the following section we illustrate the utilization of structures of capabilities providers with a few very simple examples.

5.3 EXAMPLES

In this section we present some simple examples to illustrate the application of structures of capability providers.

5.3.1 Mobile robots

Let us consider the task of carrying a weight, to be performed by a mobile robot. Let us assume that a mobile robot is characterized by the following attributes: speed (s), autonomy (a), cost per

5.3 EXAMPLES

Table 5.1 Values of attributes of robots.

Robot	r_1	r_2	r_3	r_4
Speed s (m min^{-1})	5	10	3	5
Autonomy a (m)	100	140	80	150
Cost c ($\$ \, m^{-1}$)	6	7	3	10
Maximum weight w (kg)	20	10	30	40

distance (c) and maximum weight that can be carried (w). The actions that can be performed by a robot are *pick weight*, *carry weight* and *drop weight*, limited to the values of its attributes. The only goal of a robot is to move a weight from an origin to a destination as fast as possible, whenever asked to do so.

Let us assume we have four robots, identified as r_1, r_2, r_3 and r_4. The values determined for the attributes of each robot are shown in Table 5.1.

We also have a special agent called *broker*, which undertakes the task of picking up equipment at a station and delivering it to another station. The distance between the two stations – henceforth called *origin* and *destination* – and the weight of the equipment are known. The robots are not assumed to be at the origin when the task is proposed to the broker, and the initial distances between each robot and the origin are also known.

We assume that the weight of the equipment is 50 kg and that the distance between origin and destination is 100 m. We also assume the following initial distances between each robot and the origin: $r_1 = 20$ m, $r_2 = 10$ m, $r_3 = 30$ m, $r_4 = 10$ m. This situation is depicted in Figure 5.6.

Two alternative metrics to evaluate the quality of service of each robot can be envisaged: speed and cost of service. We consider each one in turn.

Considering speed, the broker can estimate the quality of service of robots by looking at the values of the corresponding

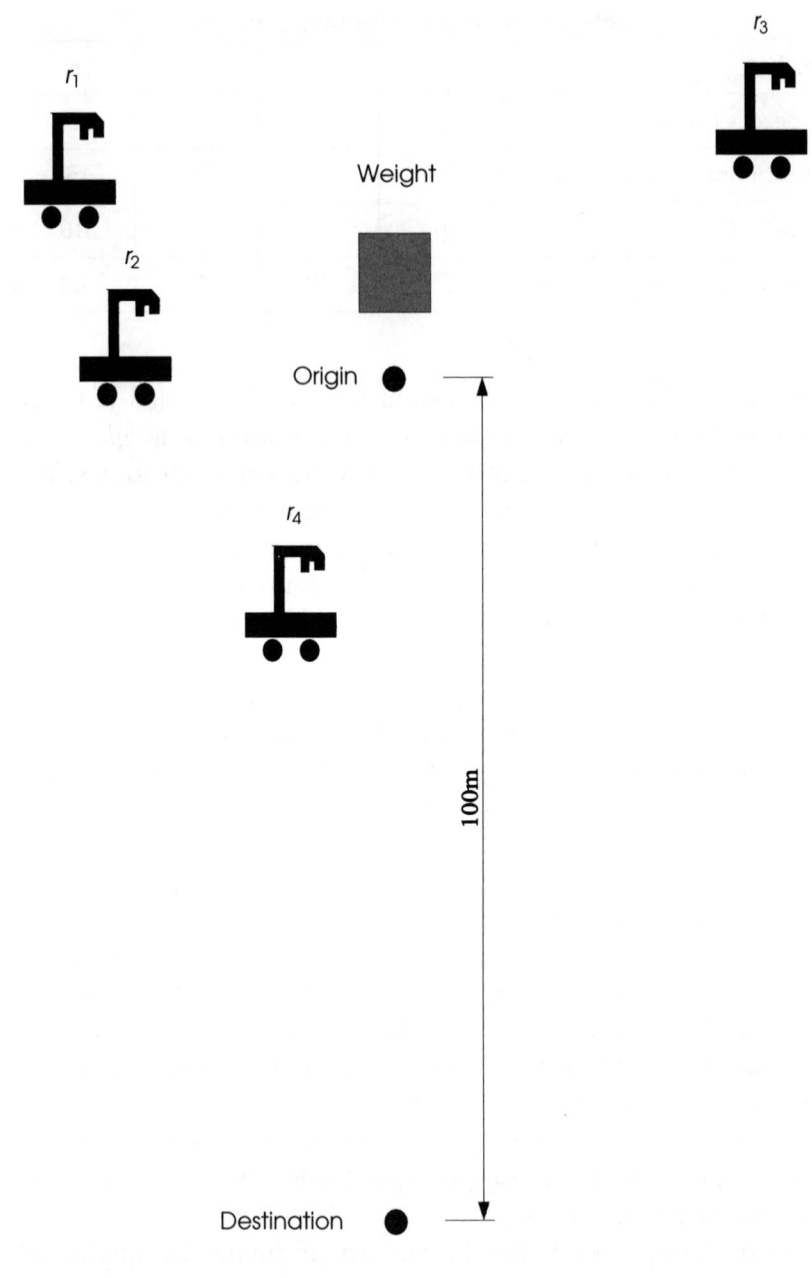

Figure 5.6 Mobile robots.

5.3 EXAMPLES

attribute, regardless of their initial position. This leads to the following partial order of robots (Figure 5.7):

- $r_1 \subseteq r_2$;
- $r_4 \subseteq r_2$;
- $r_3 \subseteq r_1$;
- $r_3 \subseteq r_4$.

However, none of these robots is capable of lifting 50 kg by itself. Hence, they must form larger agencies to execute this task. We assume the following rules for the aggregation of robots. If two or more robots become components of an agency, then:

- the speed of the agency becomes the minimum speed of its components;
- the autonomy of the agency also becomes the minumum autonomy of its components;
- the cost per distance of the agency becomes the sum of the costs of its components; and
- the maximum weight the agency can carry also becomes the sum of the weights each component can carry.

Considering all possible agencies that can be built from these four robots and selecting from them only those that have some chance of being able to undertake this task (i.e., those agencies with autonomy ≥ 100 m and capable of lifting weights ≥ 50 kg), we have the relevant agencies shown in Table 5.2.

Using the same criteria to order these agencies, we see they are not comparable with each other. Let us assume that all three

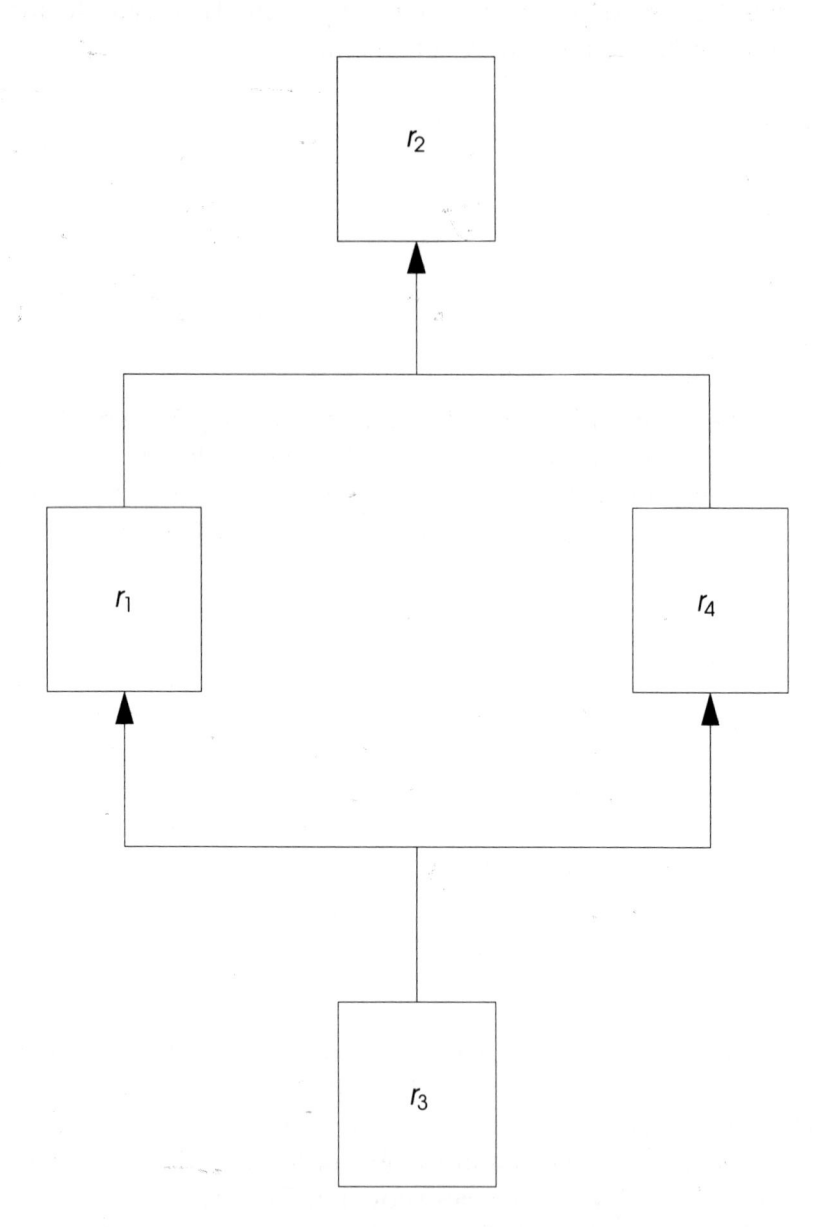

Figure 5.7 Partial order of robots by speed (estimated quality of service).

5.3 EXAMPLES

Table 5.2 Relevant agencies.

Agency	$r_1 + r_4$	$r_2 + r_4$	$r_1 + r_2 + r_4$
Speed s (m min^{-1})	5	5	5
Autonomy a (m)	100	140	100
Cost c ($\$$ m^{-1})	16	17	23
Maximum weight w (kg)	60	50	70

agencies are available to undertake this task. According to our proposed model, the broker should select the first (according to the partial order in use) agency capable of executing the task with quality compatible with the one estimated for that agency. In our case, this means taking the equipment from the origin to the destination in 20 minutes (since the distance between the origin and destination is 100 m and the agencies run at 5 m min^{-1}). Since the robots are not initially at the origin, none of these agencies can execute the task at this estimated time. The time necessary for each component of these agencies to reach the origin is:

- $r_1 = \dfrac{20\,\text{m}}{5\,\text{m min}^{-1}} = 4$ minutes;

- $r_2 = \dfrac{10\,\text{m}}{10\,\text{m min}^{-1}} = 1$ minute;

- $r_4 = \dfrac{10\,\text{m}}{5\,\text{m min}^{-1}} = 2$ minutes.

The real time necessary for each agency to execute the task is, therefore:

- $r_1 + r_4 = 20 + max\{4, 2\} = 24$ minutes;

- $r_2 + r_4 = 20 + max\{1, 2\} = 22$ minutes;

- $r_1 + r_2 + r_4 = 20 + max\{4, 1, 2\} = 24$ minutes.

The selected agency, in this case, should be $r_2 + r_4$.

Now, considering the cost to order the agencies, we have that $r_1 + r_2 + r_4 \subseteq r_2 + r_4 \subseteq r_1 + r_4$, since the estimated quality of service is now given by:

- $r_1 + r_2 + r_4$: $23\,\$\,m^{-1}$ (highest – i.e., "worst" – cost);
- $r_2 + r_4$: $17\,\$\,m^{-1}$;
- $r_1 + r_4$: $16\,\$m^{-1}$ (lowest – i.e., "best" – cost).

In other words, the estimated cost for the task is, for each agency:

- $r_1 + r_2 + r_4$: \$2,300;
- $r_2 + r_4$: \$1,700;
- $r_1 + r_4$: \$1,600.

This suggests that agency $r_1 + r_4$ should be selected.

Again, since the robots are not initially at the origin, the real cost of transportation is higher. According to the model we propose, some additional considerations should be made before the final decision is reached.

The real cost for agency $r_1 + r_4$ is \$1,600 plus the cost for agents r_1 and r_4 to reach the origin. This amounts to $\$1{,}600 + 20 \times \$6 + 10 \times \$10 = \$1{,}820$. Since this value is greater than the estimated value for the next agency in the order (i.e., agency $r_2 + r_4$, with an estimated cost of \$1,700), the real cost for that agency must be calculated. The real cost for $r_2 + r_4$ is $\$1{,}700 + 10 \times \$7 + 10 \times \$10 = \$1{,}870$, which surpasses the real cost for $r_1 + r_4$. Since the real cost for $r_1 + r_4$ is greater than the estimated cost for $r_1 + r_2 + r_4$, the real cost for that agency must not be taken into account. Hence, in this particular case, the selected agency is indeed $r_1 + r_4$.

This a very simple example in which we show a structure of capability providers in use. Slightly more convoluted examples can be built (e.g., by considering large amounts of equipment to be transported and asynchronous tasks).

5.3.2 Conference speakers

Let us now consider a conference organizer who is looking at likely candidates to speak at an invited talk and run a mini-course at the conference. The conference organizer knows three potential candidates to execute the task, each with a certain reputation as a specialist and with different availabilities for the period of the conference (Table 5.3).

Let us consider that the task can be refined into two subtasks (presenting an invited talk and presenting a mini-course), and that this can be done by sequential decomposition or by parallel decomposition. Sequential decomposition admits a single invited speaker as the solution, while parallel decomposition requires at least two speakers.

If the conference schedule permits the invited talk and the mini-course to be disjoint in time, the best decision is to invite speaker s_1 for both. If the conference schedule is such that the refinement of this task must be by parallel decomposition, then the initial choice would be $s_1 + s_2$. However, since s_2 is not available, this choice would have to be revised to $s_1 + s_3$.

Table 5.3 Values of attributes of conference speakers.

Speaker	s_1	s_2	s_3
Reputation	High	High	Medium
Available	Yes	No	Yes

This even simpler example illustrates the possible consequences of refining a task. This shows that structures of capability providers can also be easily employed for human agencies.

5.3.3 Other examples

Many other examples of how to use structures of capability providers can be devised: selection of software products to accomplish specific tasks, repair service providers for specific industrial equipment, banking and financial services, airline tickets, etc.

5.4 ASSESSING KNOWLEDGE COORDINATION

We propose that knowledge coordination occur on the basis of management of structures of capability providers. This can be the main tool to discipline the way agencies are called upon to perform tasks within an organization.

Hence, structures of capability providers can be an important tool for knowledge management. The delegation and execution of tasks within an organization can be controlled via the statement of specific partial orderings among capability providers.

The global organization of structures of capability providers can be the object of analysis and improvement. A knowledge manager can, for example, run *simulated traces of tasks delegation*, based on which many different analyses can be done. The results of these analyses can guide the empirical adjustment of the existing structures of capability providers, leading the behaviour of the organization as a whole to greater levels of efficiency and effectiveness, in the sense of Barnard (1938). Some possible analyses that can be performed are detailed below.

5.4 ASSESSING KNOWLEDGE COORDINATION

5.4.1 Minimize $\frac{\text{(delegations)}}{\text{(task)}}$

A common anomaly found in bureaucratic organizations is the excessive delegation of tasks, which amounts to a great deal of energy spent in translating and communicating tasks, but little energy spent in actually executing them.

This anomalous behaviour can be corrected by ensuring that in most structures of capability providers, the holder of the structure also happens to be a capability provider in the structure, and that in most partial orders the holder of the structure is high-ranked.

5.4.2 Minimize $\frac{\text{(agents)}}{\text{(task)}}$

It can happen that a small number of agencies are required to execute a task, but that each agency has a large number of components. The most unwanted consequence of this is that tasks spread through large parts of the organization very frequently, thus keeping too many agents busy most of the time and possibly clogging up the potential to undertake more tasks.

Structures of capability providers can be restructured in this case, so that the number of alternative agents used to execute each task is limited.

5.4.3 Maximize probability of cross-delegation of tasks

Probable cross-delegation of tasks can occur if the holder of a structure of capability providers S appears as a high-ranked provider in the structures of the agencies that are high-ranked holders in S.

By restructuring the structures of capability providers to maximize the probability of cross-delegation of tasks, a manager stimulates the contact between specified pairs of agencies. This is how we envisage utilization of this conceptual tool to manage the formation of communities of practice.

REFERENCES

Barnard, C. I. (1938) *The Functions of the Executive*, Harvard University Press.

Davenport, T. H. and Prusak, L. (1998) *Working Knowledge*, HBS Press.

Fensel, D. (2001) *Ontologies: A Silver Bullet for Knowledge Management and Electronic Commerce*, Springer-Verlag.

Fensel, D., van Harmelen, F., Horrocks, I., McGuinness, D. L. and Patel-Schneider, P. F. (2001) "OIL: An ontology infrastructure for the semantic web," *IEEE Intelligent Systems*, **16**(2), 38–45.

Liebowitz, J. (2001) *Knowledge Management: Learning from Knowledge Engineering*, CRC Press.

Tiwana, A. (1999) *The Knowledge Management Toolkit: Practical Techniques for Building a Knowledge Management System*, Prentice Hall.

6

Conclusion

Try again. Fail again. Fail better. – S. Beckett

Evidently, this work does not have a "conclusion". It is deemed an eternal work in progress, as every concept treated here unfolds to a myriad of interesting and thought-provoking themes to be discussed. Nevertheless, we considered it would be appropriate to close the book with some general observations about our writings – some you have already seen in this book and others we are preparing at the time of publication of this book.

We started this book with some observations on how the concept of knowledge has presented itself to managers over the years, emphasizing two issues in particular:

- Knowledge as a concept is complex, subtle and multifaceted. Quite frequently we are satisfied with a partial understanding of knowledge that serves our purposes. The problem is that the partial views that coexist in different texts related to "knowledge" – management, engineering, society, etc. – abound. These different partial views are frequently taken for granted, as if there were some consensus about the meaning of knowledge. We hope to have alerted the reader to the coexistence of these many partial conceptualizations of knowledge in different texts. We have analysed the most frequently found concepts, highlighting the understanding of knowledge as the capability to act, which seemed to us the most useful conceptualization of knowledge for present management practices.

- Management of knowledge can be based on two fundamental strategies:

    - the codification strategy, whose aim is the extraction of knowledge from agencies in such a way that those agencies become dispensable – or at least replaceable; and

    - the personalization strategy, founded on the principle that significant portions of knowledge cannot be taken from agencies that hold them, hence some agencies are fundamentally irreplaceable.

Codification and personalization have been empirically observed to exclude each other. We have suggested throughout this book that this must not be so and that *some* significant portions of knowledge are not allowed to be taken from their originating agencies, while other portions are. The codification strategy for knowledge management can be effective and efficient (in Barnard's sense) when it is applicable. For those pieces of knowledge that defy codification we have proposed alternative information structures to support knowledge management.

More specifically, we have devised a very simple, yet effective conceptual tool for knowledge management – the structure of capability providers – which we believe can be very useful for an encompassing knowledge management that is based on both codification and personalization.

CONCLUSION

We have complained throughout the text about the misuse of the term *ontology*, which in our view has a broad meaning, one that was well established before being borrowed by information technologists to convey some other meaning (although related to the original one). We have therefore tagged the concept proposed by information technologists as *artificial ontologies*.

Nevertheless, we have acceded to the view of knowledge as the conduit of action. However, we want to make it clear that we also do not agree with this simplification of the concept of knowledge. Indeed, in our view this is a poor interpretation of a wide-ranging term that should be reserved for much nobler concepts.

We have retained this simplistic understanding of knowledge because it seemed useful to introduce our views about knowledge coordination in business organizations. We believe this could be important and interesting enough to warrant a book of its own, for two reasons:

- From a very practical point of view, because there are a lot of managers and information technologists working out there (we would be very pleased to hear from you, dear reader, if you are one of them: [Flavio – fcs@ime.usp.br, Jaume – agusti@iiia.csic.es]), we consider that our analyses of current practices and technologies, as well as our small contributions to the themes treated in this book, could be useful for those professionals.

- From a philosophical point of view, we believe that the surge of interest in knowledge – what it is, where it comes from, where it goes, where it resides and what it does to all sorts of entities as it goes by – may indicate an important change in our world, economies, societies and organizations. When it comes down to business organizations (from a Western capitalist viewpoint), knowledge is a close relative of action and agencies. Therefore, we consider the work organized and presented here could be a good starting point to discuss a broader understanding of knowledge and how it permeates and shapes the world in which we live.

The essence of what we have proposed here is that organizations are composed of agencies, which ultimately are formed by *agents*. Every action originates from agents, every capability is formed by those capabilities found in agents, likewise for goals and motivations. Hence, to understand organizations and agencies, and/or to shape, control and guide their behaviour, we should look at the agents that form these agencies and organizations. General rules, laws, regulations, control procedures, etc. should always be regarded as means to the expression of agents who hold the active role in organizations, and never the other way round: organizations result from the actions of agents, and agents are not the consequence of the structuring of organizations.

This view underlies what we have presented here, and we believe that the limited view of knowledge as the conduit of action we have employed is useful to illustrate this view. We do not claim, however, that this understanding of knowledge suffices to provide a convincing argumentation that gives room for such conclusions.

We need to "take the bull by the horns" if we want to build such an argumentation.

In the companion book (under preparation) to this one, we undertake this task. We also note that the European, post-Renaissance model of the world is probably not enough to clinch it, so we are studying and analysing the understanding of knowledge and organizations found in ancient Europe, as well as in some non-European traditions (e.g., native American Indian and eastern traditions – Indian, Chinese, etc.).

The companion book will follow similar lines to the present one: it will be practical and expound explicit technical proposals.

CONCLUSION

It will, nevertheless, be more philosophical and speculative than this one.

When we unshackle knowledge from the burden of expressing itself in actions, we are able to see the more subtle expressions of knowledge. In the companion book we focus on knowledge itself, not on coordination or management as here. We hope with this work to contribute to the clarification of points that can be useful in building a better world for everyone. Although we do not presume that any one of our views is by itself a solution to any problem or situation we find in our world, we wholeheartedly believe that by insisting on the discussion of these issues our work can be of help.

For the moment, dear reader, we thank you for sticking with us to reach these final lines. We hope to have entertained you and that this reading may have been useful to you.

Jaume and Flávio
January 2003

Bibliography

Apt, K. R. (1994) "Logic programming," in J. van Leeuwen (ed.), *Handbook of Theoretical Computer Science*, Elsevier–MIT Press.

Barnard, C. I. (1936) *Mind in Everyday Affairs*, University of Princeton Press (Cyrus Fogg Brackett Lecture).

Barnard, C. I. (1938) *The Functions of the Executive*, Harvard University Press.

Bernus, P. and Nemes, L. (1999) "Organisational design: Dynamically creating and sustaining integrated virtual enterprises," *Proceedings of IFAC World Congress, Beijing*.

Binney, D. (2002) "The knowledge management spectrum: The human factor," in E. Coakes, D. Willis and S. Clarke (eds), *Knowledge Management in the Sociotechnical World*, Springer-Verlag.

Carbogim, D. V. and Corrêa da Silva, F. S. (1998) "Annotated logic applications for imperfect information," *Applied Intelligence*, **9**, 163–172.

Carley, K. M. (1995) "Computational and mathematical organization theory: Perspective and directions," *Computational and Mathematical Organization Theory*, **1**(1).

Knowledge Coordination   F. S. Corrêa da Silva and J. Agustí-Cullell
© 2003 John Wiley & Sons, Ltd   ISBN: 0-470-85832-X

Cherns, A. (1987) "Principles of sociotechnical design revisited," *Human Relations*, **40**(3), 153–162.

Coakes, E. (2002) "Knowledge management: A sociotechnical perspective," in E. Coakes, D. Willis and S. Clarke (eds), *Knowledge Management in the Sociotechnical World*, Springer-Verlag.

Corrêa da Silva, F. S. and Meneses, E. X. (2001) "Integração de agentes de informação," in A. T. Martins and D. L. Borges (eds), *Anais do XXI Congresso da Sociedade Brasileira de Computação*, **3** (Jornada de Atualização em Inteligencia Artificial, Brazil) [in Portuguese].

Corrêa da Silva, F. S. and Meneses, E. X. (2002) "Expressing systems capabilities for knowledge coordination," in *Proceedings of AAMAS – Autonomous Agents and Multi-agent Systems, Italy*.

Corrêa da Silva, F. S., Vasconcelos, W. W., Agustí-Cullell, J., Robertson, D. S. and Melo, A. C. V. (1999) "Why ontologies are not enough for knowledge sharing," in *12th International Conference on Industrial and Engineering Applications of Artificial Intelligence and Expert Systems, Egypt*.

Corrêa da Silva, F. S., Vasconcelos, W. W., Robertson, D. S., Brilhante, V., Melo, A. C. V., Finger, M. and Agustí-Cullell, J. (2001) "On the insufficiency of ontologies: Problems in knowledge sharing and alternative solutions," *Knowledge Based Systems*, **14**(7).

Cummins, R. and Pollock, J. (eds) (1991) *Philosophy and AI: Essays at the Interface*, MIT Press.

Daft, R. L. (2001) *Organization Theory and Design*, Thomson, 2001 (7th edition).

Davenport, T. H. and Prusak, L. (1998) *Working Knowledge*, HBS Press.

Delgrande, J. P. and Mylopoulos, J. (1986) "Knowledge representation: Features of knowledge," in W. Bibel and P. Jorrand (eds), *Fundamentals of Artificial Intelligence: An Advanced Course*, Springer-Verlag (Lecture Notes in Computer Science No. 232).

Ding, Y., Fensel, D., Klein, M. and Omelayenko, B. (2002) "The semantic web: Yet another hip?," *Data and Knowledge Engineering*, **41**, 205–227.

d'Inverno, M. and Luck, M. (2001) *Understanding Agent Systems*, Springer-Verlag.

d'Inverno, M., Fisher, M., Lomuscio, A., Luck, M., de Rijke, M., Ryan, M. and Wooldridge, M. (1997) "Formalisms for multi-agent systems," *The Knowledge Engineering Review*, **12**(3), 315–321.

Drucker, P. F. (1988) "The coming of the new organization," *Harvard Business Review*, **66**(1), January–February, 45–53.

Fagin, R., Halpern, J. Y., Moses, Y. and Vardi, M. Y. (1995) *Reasoning about Knowledge*, MIT Press.

Fayol, H. (1916) "Administration industrielle et générale," *Bulletin de la Société de l'Industrie Minerale* (Dunod) [in French].

Fensel, D. (2001) *Ontologies: A Silver Bullet for Knowledge Management and Electronic Commerce*, Springer-Verlag.

Fensel, D., van Harmelen, F., Horrocks, I., McGuinness, D. L. and Patel-Schneider, P. F. (2001) "OIL: An ontology infrastructure for the semantic web," *IEEE Intelligent Systems*, **16**(2), 38–45.

Fischer, G. and Ostwald, J. (2001) "Knowledge management: Problems, promises, realities, and challenges," *IEEE Intelligent Systems*, 60–72.

Fisher, M., Muller, J., Schroeder, M., Stanford, G. and Wagner, G. (1997) "Methodological foundations for agent based systems," *The Knowledge Engineering Review*, **12**(3), 323–329.

Fleury, A. and Fleury, M. T. L. (1997) *Aprendizagem e Inovação Organizacional*, Atlas [in Portuguese].

Fleury, A. and Vargas, N. (eds) (1983) *Organização do Trabalho*, Atlas [in Portuguese].

Fox, M. S., Barbuceanu, M., Gruninger, M. and Lin, J. (1998) "An organisational ontology for enterprise modeling," in M. J.

Prietula, K. M. Carley and L. Gasser (eds), *Simulating Organizations: Computational Models of Institutions and Groups*, AAAI/MIT Press.

Garvin, D. A. (1993) "Building a learning organization," *Harvard Business Review*, July–August, 81–91.

Geerts, G. L. and McCarthy, W. E. (1999) "An accounting object infrastructure for knowledge-based enterprise models," *IEEE Intelligent Systems*, July/August, 1–6.

Gettier, E. L. (1963) "Is justified true belief knowledge?," *Analysis*, **23**(6), 121–123.

Gruber, T. R. (1993) "A translation approach to portable ontologies," *Knowledge Acquisition*, **5**(2), 199–220.

Hansen, M. T., Nohria, N. and Tierney, T. (1999) "What's your strategy for managing knowledge?," *Harvard Business Review*, **106**, March–April.

Haugen, R. and McCarthy, W. E. (2000) "REA, a semantic model for Internet supply chain collaboration," *OOPSLA – ACM Conference on Object-oriented Programming, Systems, Languages, and Applications – Business Objects and Component Design and Implementation Workshop VI: Enterprise Application Integration*, Minneapolis, MN.

Hempel, C. G. (1958) "The theoretician's dilemma," in H. Feigl, M. Scriven and G. Maxwell (eds), *Minnesota Studies in the Philosophy of Science II*, University of Minnesota Press.

Howard, D. (1996) *Discussion on Roles*, BPR List.

Kalfoglou, Y., Menzies, T., Althoff, K. and Motta, E. (2000) "Meta-knowledge in systems design: Panacea ... or undelivered promise?," *The Knowledge Engineering Review*, **15**(4), 381–404.

Kinny, D. (2001) "Reliable agent communication – a pragmatic perspective," *New Generation Computing*, **19**, 139–156.

Klusch, M. (2001) "Information agent technology for the Internet: A survey," *Data and Knowledge Engineering*, **36**, 337–372.

Kyburg, H. (1991) "Normative and descriptive ideals," in R. Cummins and J. Pollock (eds), *Philosophy and AI: Essays at the Interface*, MIT Press.

Liebowitz, J. (2001) *Knowledge Management: Learning from Knowledge Engineering*, CRC Press.

Luck, M. and d'Inverno, M. (2001) "A conceptual framework for agent definition and development," *The Computer Journal*, **44**(1), 1–20.

Marshall, C. (1999) *Enterprise Modeling with UML: Designing Successful Software through Business Analysis*, Addison-Wesley.

McCarthy, W. E. (1982) "The REA accounting model: A generalised framework for accounting systems in a shared data environment," *The Accounting Review*, 554–578.

McCarthy, W. E. and Geerts, G. L. (1999) "Modeling business enterprises as value-added process hierarchies with resource-event-agent object templates," in J. Sutherland and D. Patel (eds), *Business Object Design and Implementation*, Springer-Verlag (pp. 113–128).

Mendelson, E. (1987) *Introduction to Mathematical Logic*, Wadsworth and Brooks/Cole (3rd edition).

Muller, J. P. (1998) "Architectures and applications of intelligent agents: A survey," *The Knowledge Engineering Review*, **13**(4), 353–380.

Nonaka, I. (1991) "The knowledge creating company," *Harvard Business Review*, November–December, 96–104.

Noy, N. F., Fergerson, R. W. and Musen, M. A. (2000) "The knowledge model of Protégé-2000: Combining interoperability and flexibility," *2nd International Conference on Knowledge Engineering and Knowledge Management (EKAW 2000)*, Juan-les-Pins, France.

Noy, N. F. and Musen, M. A. (2000) *PROMPT: Algorithm and Tool for Automated Ontology Merging and Alignment*, AAAI/IAAI – 2000, Austin, TX.

Plato "Theaetetus," in E. Hamilton and H. Cairns (eds), *The Collected Dialogues*, Princeton University Press (16th edition).

Preece, A., Flett, A., Sleeman, D., Curry, D., Meany, N. and Perry, P. (2001) "Better knowledge management through knowledge engineering," *IEEE Intelligent Systems*, 36–42.

Prietula, M. J., Carley, K. M. and Gasser, L. (eds) (1998) *Simulating Organizations: Computational Models of Institutions and Groups*, AAAI/MIT Press.

Ramsey, F. (1931) "Theories," in *The Foundations of Mathematics and Other Logical Essays*, Littlefield, Adams & Co.

Robbins, S. P. (1990) *Organization Theory: Structure, Design and Applications*, Prentice Hall.

Robertson, D. S., Agustí-Cullell, J., Corrêa da Silva, F. S., Vasconcelos, W. W. and Melo, A. C. V. (2000) "A lightweight capability communication mechanism," *13th. International Conference on Industrial and Engineering Applications of Artificial Intelligence and Expert Systems*, New Orleans.

Robinson, J. A. (1965) "A machine-oriented logic based on the resolution principle," *Journal of the ACM*, **12**(1).

Russell, S. and Norvig, P. (1995) *Artificial Intelligence – A Modern Approach*, Prentice Hall.

Scholtz, V. (2002) "Managing knowledge in a knowledge business," in E. Coakes, D. Willis and S. Clarke (eds), *Knowledge Management in the Sociotechnical World*, Springer-Verlag.

Senge, P. (1990) *The Fifth Discipline*, Currency/Doubleday.

Senge, P., Kleiner, A., Roberts, Ch., Ross, R. and Smith, B. (1994) *The Fifth Discipline Field Book: Strategies and Tools for Building A Learning Organization*, Currency/Doubleday.

Senge, P., Kleiner, A., Roberts, Ch., Ross, R., Roth, G. and Smith, B. (1999) *The Dance of Change: The Challenges of Sustaining Momentum in Learning Organizations*, Currency/Doubleday.

Shoenfield, J. (1967) *Mathematical Logic*, Addison-Wesley.

Silva, R. O. (2001) *Teorías da Administração*, Thomson [in Portuguese].

Starkey, K. (ed.) (1992) *How Organizations Learn*, Thomson Business Press.

Sowa, J. F. (2000) *Knowledge Representation – Logical, Philosophical and Computational Foundations*, Brooks/Cole.

Sterling, L. and Shapiro, E. (1986) *The Art of Prolog*, MIT Press.

Tarski, A. (1955) A lattice-theoretic fix-point theorem and its applications. *Pacific Journal of Mathematics*, **5**, 285–309.

Taylor, F. W. (1911) *Principles of Scientific Management*, Harper & Row.

Terra, J. C. C. (2000) "Gestão de conhecimento: O grande desafio empresarial," *Negocio* [in Portuguese].

Tiwana, A. (1999) *The Knowledge Management Toolkit: Practical Techniques for Building a Knowledge Management System*, Prentice Hall.

Tuomela, R. (1973) *Theoretical Concepts*, Springer-Verlag.

Turi, D. (1989) *Logic Programs with Negation: Classes, Models, Interpreters*, Center for Mathematics and Computer Science (Technical Report CS-R8943).

Uschold, M. (1998) "Where are the killer apps?," *Workshop on Applications of Ontologies and Problem Solving Methods, ECA 1998, Brighton, UK*.

Uschold, M., Jasper, R. and Clark, P. (1999) "Three approaches for knowledge sharing: A comparative analysis," *12th Workshop on Knowledge Acquisition, Modeling and Management, Banff, Alberta, Canada*.

Verharen, E. M. (1997) *A Language-action Perspective on the Design of Cooperative Information Agents*, The InfoLab, Tilburg University.

Watt, D. (1990) *Programming Language Concepts and Paradigms*, Prentice Hall International Series in Computer Science.

Weber, M. (1947) *The Theory of Social and Economic Organisation*, Free Press.

Winograd, T. (1997) "The design of interaction," in P. J. Denning and R. M. Metcalfe (eds), *Beyond Calculation: The Next Fifty Years of Computing*, Springer-Verlag.

Index

Action, 84
Agency, 4
Agustí-Cullell, J., 64, 65, 132, 164, 168
Althoff, K., 133, 166
Apt, K. R., 70, 94, 163
Attribute, 69
   instance, 69
   value, 69

Barbuceanu, M., 14, 166
Barnard, C. I., 3, 9, 12, 13, 14, 19, 20, 21, 28–34, 39, 43, 62, 63, 64, 90, 91, 93, 142, 154, 156, 158, 163
Belief, 50
Bernus, P., 94, 163
Binney, D., 14, 163
Brilhante, V., 64, 132, 164

Carbogim, D., 64, 163
Carley, K. M., 94, 163
Cherns, A., 35, 64, 164
Clausal theory, 69
Coakes, E., 64, 164
Codification, 9, 10, 14, 18, 63, 88, 92, 93, 94, 106, 117, 129, 130
Community of practice, 62, 136
Coordination, 4

Corrêa da Silva, F. S., 64, 65, 121, 131, 132, 163, 164, 168
Cummins, R., 11, 14, 164
Curry, D., 65, 168

Daft, R. L., 14, 64, 164
Data, 44
Davenport, T. H., 14, 64, 132, 156, 164
Delgrande, J. P., 64, 164
Ding, Y., 132, 165
Drucker, P. F., 14, 64, 165

Effectiveness, 33
Efficiency, 33
Engineering, 7
Existence, 100
   absolute, 100
   social, 100
   subjective, 100

Fagin, R., 64, 165
Fayol, H., 64, 165
Fensel, D., 132, 165
Fergerson, R. W., 133, 167
Finger, M., 64, 132, 164
Fischer, G., 65, 165
Fisher, M., 94, 95, 165

Flett, A., 65, 168
Fleury, A., 65, 165
Fleury, M. T. L., 65, 165
Ford, H., 23, 24, 26, 27, 92
Fox, M. S., 14, 165

Garvin, D. A., 132, 166
Gasser, L., 15, 95, 168
Geerts, G. L., 132, 133, 166, 167
Gettier, E. L., 65, 166
Goal, 84
Gruber, T. R., 132, 166
Gruninger, M., 14, 165

Halpern, J. Y., 64, 165
Hansen, M. T., 15, 65, 95, 132, 166
Harmelen, F. van, 132, 165
Haugen, R., 132, 166
Hempel, C. G., 65, 166
Horn clause, 72
Horrocks, I., 132, 165
Howard, D., 15, 166

Information, 44
Interpretation, 44–49
 complete, 45
 incomplete, 45
 partial, 45
 total, 45
Inverno, M., 95, 165, 167

Justification, 52

Kalfoglou, Y., 133, 166
Kinny, D., 133, 166
Klein, M., 132, 165
Kleiner, A., 65, 168
Klusch, M., 133, 166
Knowledge base, 9, 10, 18, 39, 62, 63, 88, 94, 104, 105, 108, 109, 114
Knowledge management, 1, 2, 9–12, 18–21, 28, 34, 35, 38, 39, 62, 63, 88–90, 92, 94, 97, 101, 106, 117, 130, 131, 135, 137, 154, 158
Knowledge map, 92, 137, 138
Kyburg, H., 15, 167

Liebowitz, J., 65, 156, 167
Lin, J., 14, 165
Logic program, 73
Logic programming, 69–84
Lomuscio, A., 95, 165
Luck, M., 95, 165, 167

Management, 4
Marshall, C., 15, 167
McCarthy, W. E., 106, 132, 133, 166, 167
McGuinness, D. L., 132, 165
Meany, N., 65, 168
Melo, A. C. V., 65, 132, 164, 168
Mendelson, E., 95, 167
Meneses, E. X., 131, 132, 164
Menzies, T., 133, 166
Moses, Y., 64, 165
Motivation, 84–88, 90, 92
Motta, E., 133, 166
Muller, J., 94, 95, 165, 167
Musen, M. A., 133, 167
Mylopoulos, J., 64, 164

Nemes, L., 94, 163
Nohria, N., 15, 65, 95, 132, 166
Nonaka, I., 133, 167
Norvig, P., 95, 168
Noy, N., 133, 167

Omelayenko, B., 132, 165
Organization, 3, 4
Ostwald, J., 65, 165

Patel–Schneider, P. F., 132, 165
Perception, 86
Perry, P., 65, 168

Personalization, 9, 10, 14, 18, 62, 63, 88, 92–94, 131, 135, 137, 158
Plan, 87
Plato, 65, 168
Pollock, J., 11, 14, 164
Preece, A., 65, 168
Prietula, M. J., 94, 163
Protégé, 108–113, 122
Prusak, L., 14, 64, 132, 156, 164

Ramsey, F., 65, 168
Referent, 45
Rijke, M., 95, 165
Robbins, S. P., 15, 168
Roberts, Ch., 65, 168
Robertson, D. S., 64, 65, 132, 164, 168
Robinson, J. A., 95, 168
Ross, R., 65, 168
Roth, G., 65, 168
Russell, S., 95, 168
Ryan, M., 95, 165

Scholtz, V., 15, 168
Schroeder, M., 94, 95, 165
Senge, P., 15, 65, 133, 168
Shoenfield, J., 95, 167
Silva, R. O., 65, 169
Sleeman, D., 65, 168
Smith, B., 65, 168
Society, 8
Sowa, J. F., 15, 169

Stanford, G., 94, 95, 165
Starkey, K., 133, 169
State, 84

Taylor, F. W., 21–23, 66, 169
Terra, J. C. C., 66, 169
Tierney, T., 15, 65, 95, 132, 166
Tiwana, A., 15, 66, 156, 169
Truth, 51
  absolute, 51
  common, 51
  distributed, 51
  group, 51
  social, 51
  subjective, 51
Tuomela, R., 66, 169

Uschold, M., 133, 169

Vardi, M. Y., 64, 165
Vargas, N., 65, 165
Vasconcelos, W., 64, 65, 132, 164, 168
Verharen, E. M., 15, 169

Wagner, G., 94, 95, 165
Watt, D., 15, 169
Weber, M., 26, 27, 66, 169
Winograd, T., 15, 169
Wooldridge, M., 95, 165